Report on Planet Three
and Other Speculations

Report on Planet Three

and Other Speculations

Arthur C. *Clarke*

HARPER & ROW, PUBLISHERS

1817

NEW YORK, EVANSTON
SAN FRANCISCO, LONDON

Portions of this work have previously appeared in *Boys' Life, Cavalier, Fantasy & Science Fiction, Glamour, Holiday, New York Magazine, Playboy, Readers' Digest,* and *Why Not.*

The quotation from G. K. Chesterton's "The Song of Quoodle" appears in *The Collected Poems of G. K. Chesterton,* published by Dodd, Mead & Company.

FIRST EDITION

STANDARD BOOK NUMBER 06–010793–6

LIBRARY OF CONGRESS CATALOG CARD NUMBER: 74–156515

In accordance with the terms of the Clarke-Asimov Treaty, the second-best science writer dedicates this book to the second-best science-fiction writer.

Contents

III. *The Technological Future*

IV. *Frontiers of Science*

V. *Son of Dr. Strangelove, Etc. . . .*

Preface

My first volume of speculative essays, *The Challenge of the Spaceship,* was published in 1959—exactly one decade before Neil Armstrong set foot on the Moon—and has now been out of print for some years. During the sixties two other collections appeared, *Profiles of the Future* (1963) and *Voices from the Sky* (1965), and these are still available to sufficiently determined readers.

In the meantime, much has happened, and I have had later thoughts on a number of subjects. Yet the events that culminated in July 1969 at Tranquility Base have not made these speculations obsolete; indeed, many of them are now even more timely than when they first appeared.

A number of the following pieces are reprinted from *The Challenge of the Spaceship;* with very few exceptions, the others have never appeared in book form. For convenience, they have been divided into five categories: Talking of Space; Outward from Earth; The Technological Future; Frontiers of Science; and Son of Dr. Strangelove, Etc. . . . However, there is no particular order in which they need be read, and even old-fashioned pre-McLuhan readers are welcome to abandon linearity if they wish.

In the closing years of man's last earthbound era, I ended the precursor of this book with the words:

Across the gulf of centuries, the blind smile of Homer is turned upon our age. Along the echoing corridors of time, the roar of the

rockets merges now with the creak of the wind-taut rigging. For somewhere in the world today, still unconscious of his destiny, walks the
boy who will be the first Odysseus of the Age of Space. . . .

Who could have dreamed, back in 1959, that the "boy" was then
already nearing his thirtieth birthday? But there will be other Odys-
seys to come. . . .

Colombo
January 1971

I. *Talking of Space*

Report on Planet Three

1

[*The following document, which has just been deciphered for the Interplanetary Archaeological Commission, is one of the most remarkable that has yet been discovered on Mars, and throws a vivid light upon the scientific knowledge and mental processes of our vanished neighbors. It dates from the Late Uranium (i.e., final) Age of the Martian civilization, and thus was written little more than a thousand years before the birth of Christ.*

The translation is believed to be reasonably accurate, though a few conjectural passages have been indicated. Where necessary, Martian terms and units have been converted into their terrestrial equivalents for ease of understanding.—TRANSLATOR.]

The recent close approach of the planet Earth has once again revived speculations about the possibility of life upon our nearest neighbor in space. This is a question which has been debated for centuries, without conclusive results. In the last few years, however, the development of new astronomical instruments has given us much more accurate information about the other planets. Though we cannot yet confirm or deny the existence of terrestrial life, we now have much more precise knowledge of conditions on Earth and can base our discussions on a firm scientific foundation.

One of the most tantalizing things about Earth is that we cannot see it when it is closest, since it is then between us and the Sun and its dark side is therefore turned toward us. We have to wait until it is a morning or evening star, and thus a hundred million or more miles away from us, before we can see much of its illuminated surface. In the telescope, it then appears as a brilliant crescent, with its single giant Moon hanging beside it. The contrast in color between the two bodies is striking; the Moon is a pure silvery-white, but the Earth is a sickly blue-green. [*The exact force of the adjective is uncertain; it is definitely unflattering. "Hideous" and "virulent" have been suggested as alternatives.*—TRANSLATOR.]

As the Earth turns on its axis—its day is just half an hour shorter than ours—different areas of the planet swing out of darkness and appear on the illuminated crescent. By carrying out observations over a period of weeks, it is possible to construct maps of the entire surface, and these have revealed the astonishing fact that *more than two-thirds of the planet Earth is covered with liquid.*

Despite the violent controversy that has raged over this matter for some centuries, there is no longer any reasonable doubt that this liquid is water. Rare though water now is upon Mars, we have good evidence that in the remote past much of our planet was submerged beneath vast quantities of this peculiar compound; it appears, therefore, that Earth is in a state corresponding to our own world several billion years ago. We have no way of telling how deep the terrestrial "oceans"—to give them their scientific name—may be, but some astronomers have suggested that they are as much as a thousand feet in thickness.

The planet also has a very much more abundant atmosphere than ours; calculations indicate that it is at least ten times as dense. Until quite recently, we had no way of guessing the composition of that atmosphere, but the spectroscope has now solved this problem—with surprising results. The thick gaseous envelope surrounding the Earth contains large amounts of the poisonous and very reactive element oxygen, of which scarcely a trace exists in our own air. Earth's atmosphere also holds considerable quantities of nitrogen and water vapor, which forms huge clouds, often persisting for many days and obscuring large areas of the planet.

Being some 25 per cent nearer the Sun than Mars, Earth is at a considerably higher temperature than our world. Readings taken by thermocouples attached to our largest telescopes reveal intolerable temperatures on its Equator; at higher latitudes, however, conditions are much less extreme, and the presence of extensive icecaps at both poles indicates that temperatures there are often quite comfortable. These polar icecaps never melt completely, as do ours during the summer, so they must be of immense thickness.

As Earth is a much larger planet than Mars (having twice our diameter), its gravity is a good deal more powerful. It is, indeed, no less than three times as great, so that a 170-pound man would weigh a quarter of a ton on Earth. This high gravity must have many important consequences, not all of which we can foresee. It would rule out any large forms of life, since they would be crushed under their own weight. It is something of a paradox, however, that Earth possesses mountains far higher than any that exist on Mars; this is probably another proof that it is a young and primitive planet, whose original surface features have not yet eroded away.

Looking at these well-established facts, we can now weigh the prospects for life on Earth. It must be said at once that they appear extremely poor; however, let us be open-minded and prepared to accept even the most unlikely possibilities, as long as they do not conflict with scientific laws.

The first great objection to terrestrial life—which many experts consider conclusive—is the intensely poisonous atmosphere. The presence of such large quantities of gaseous oxygen poses a major scientific problem, which we are still far from solving. Oxygen is so reactive that it cannot normally exist in the free state; on our own planet, for example, it is combined with iron to form the beautiful red deserts that cover so much of the world. It is the absence of these areas which gives Earth its unpleasant greenish hue.

Some unknown process must be taking place on Earth which liberates immense quantities of this gas. Certain speculative writers have suggested that terrestrial life forms may actually release oxygen during the course of their metabolism. Before we dismiss this idea as being too fanciful, it is worth noting that several primitive and now extinct forms of Martian vegetation did precisely this. Nevertheless, it is very hard to believe that plants of this type can exist on Earth in the inconceivably vast quantities which would be needed to provide so much free oxygen. [*We know better, of course. All the Earth's oxygen is a by-product of vegetation; our planet's original atmosphere, like that of Mars today, was oxygen-free.*— TRANSLATOR.]

Even if we assume that creatures exist on Earth which can survive in so poisonous and chemically reactive an atmosphere, the presence of these immense amounts of oxygen has two other effects. The first is rather subtle, and has only recently been discovered by a brilliant piece of theoretical research, now fully confirmed by observations.

It appears that at a great altitude in the Earth's atmosphere— some twenty or thirty miles—the oxygen forms a gas known as ozone, containing three atoms of oxygen as compared with the normal molecule's two. This gas, though it exists in very small quantities so far from the ground, has an overwhelmingly important effect upon terrestrial conditions. It almost completely blocks the ultraviolet rays of the Sun, preventing them from reaching the surface of the planet.

This fact alone would make it impossible for the life forms we know to exist on Earth. The Sun's ultraviolet radiation, which reaches the surface of Mars almost unhindered, is essential to our well-being and provides our bodies with much of their energy. Even if we could withstand the corrosive atmosphere of Earth, we should soon perish because of this lack of vital radiation.

The second result of the high oxygen concentration is even more catastrophic. It involves a terrifying phenomenon, fortunately known only in the laboratory, which scientists have christened "fire."

Many ordinary substances, when immersed in an atmosphere like that of Earth's and heated to quite modest temperatures, begin a

violent and continuous chemical reaction which does not cease until they are completely consumed. During the process, intolerable
quantities of heat and light are generated, together with clouds of
noxious gases. Those who have witnessed this phenomenon under
controlled laboratory conditions describe it as quite awe-inspiring;
it is certainly fortunate for us that it can never occur on Mars.

Yet it must be quite common on Earth—and no possible form
of life could exist in its presence. Observations of the night side of
Earth have often revealed bright glowing areas where fire is raging;
though some students of the planet have tried, optimistically, to
explain these glows as the lights of cities, this theory must be rejected. The glowing regions are much too variable; with few exceptions, they are quite short-lived, and they are not fixed in location. [*These observations were doubtless due to forest fires and vol-
canoes—the latter unknown on Mars. It is a tragic irony of fate that
had the Martian astronomers survived a few more thousand years,
they* would *have seen the lights of man's cities. We missed each
other in time by less than a millionth of the age of our planets.*
—TRANSLATOR.]

Its dense, moisture-laden atmosphere, high gravity, and close-
ness to the Sun make Earth a world of violent climatic extremes.
Storms of unimaginable intensity have been observed raging over
vast areas of the planet, some of them accompanied by spectacular
electrical disturbances, easily detected by sensitive radio receivers
here on Mars. It is hard to believe that any form of life could
withstand these natural convulsions, from which the planet is seldom completely free.

Although the range of temperatures between the terrestrial winter and summer is not so great as on our world, this is slight compensation for other handicaps. On Mars, all mobile life forms can
easily escape the winter by migration. There are no mountains or
seas to bar the way; the small size of our world—as compared with
Earth—and the greater length of the year make such seasonable
movements a simple matter, requiring an average speed of only

some ten miles a day. There is no need for us to endure the winter, and few Martian creatures do so.

It must be quite otherwise on Earth. The sheer size of the planet, coupled with the shortness of the year (which only lasts about six of our months) means that any terrestrial beings would have to migrate at a speed of about fifty miles a day in order to escape from the rigors of winter. Even if such a rate could be achieved (and the powerful gravity makes this appear most unlikely), mountains and oceans would create insuperable barriers.

Some writers of science fiction have tried to get over this difficulty by suggesting that life forms capable of aerial locomotion may have evolved on Earth. In support of this rather far-fetched idea they argue that the dense atmosphere would make flying relatively easy; however, they gloss over the fact that the high gravity would have just the reverse effect. The conception of flying animals, though a charming one, is not taken seriously by any competent biologist.

More firmly based, however, is the theory that if any terrestrial animals exist, they will be found in the extensive oceans that cover so much of the planet. It is believed that life on our own world originally evolved in the ancient Martian seas, so there is nothing at all fantastic about this idea. In the oceans, moreover, the animals of Earth would no longer have to fight the fierce gravity of their planet. Strange though it is for us to imagine creatures that could live in water, it would seem that the seas of Earth may provide a less hostile environment than the land.

Quite recently, this interesting idea has received a setback through the work of the mathematical physicists. Earth, as is well known, has a single enormous moon, which must be one of the most conspicuous objects in its sky. It is some two hundred times the diameter of even the larger of our two satellites, and though it is at a much greater distance its attraction must produce powerful effects on the planet beneath it. In particular, what are known as "tidal forces" must cause great movements in the waters of the terrestrial oceans, making them rise and fall through distances of many feet. As a result, all low-lying areas of the Earth must be subjected to twice-daily flooding; in such conditions, it is difficult to believe that any creatures could exist either in land or sea, since the two would be constantly interchanging.

To sum up, therefore, it appears that our neighbor Earth is a forbidding world of raw, violent energies, certainly quite unfitted for any type of life which now exists on Mars. That some form of vegetation may flourish beneath that rain-burdened, storm-tossed atmosphere is quite possible; indeed, many astronomers claim to have detected color changes in certain areas which they attribute to the seasonal growth of plants.

As for animals—this is pure speculation, all the evidence being against them. If they exist at all, they must be extremely powerful and massively built to resist the high gravity, probably possessing many pairs of legs and capable only of slow movement. Their clumsy bodies must be covered with thick layers of protective armor to shield them from the many dangers they must face, such as storms, fire, and the corrosive atmosphere. In view of these facts, the question of *intelligent* life on Earth must be regarded as settled. We must resign ourselves to the idea that we are the only rational beings in the Solar System.

For those romantics who still hope for a more optimistic answer, it may not be long before Planet Three reveals its last secrets to us. Current work on rocket propulsion has shown that it is quite possible to build a spacecraft that can escape from Mars and head across the cosmic gulf toward our mysterious neighbor. Though its powerful gravity would preclude a landing (except by radio-controlled robot vehicles), we could orbit Earth at a low altitude and thus observe every detail of its surface from little more than a millionth of our present distance.

Now that we have at last released the limitless energy of the atomic nucleus, we may soon use this tremendous new power to escape the bonds of our native world. Earth and its giant satellite will be merely the first celestial bodies our future explorers will survey. Beyond them lie . . .

[Unfortunately, the manuscript ends here. The remainder has been charred beyond decipherment, apparently by the thermonuclear blast that destroyed the Imperial Library, together with the rest of Oasis City. It is a curious coincidence that the missiles which ended Martian civilization were launched at a classic moment in human history. Forty million miles away, with slightly less advanced weapons, the Greeks were storming Troy. —TRANSLATOR.]

The Men on the Moon

2

This essay was written for *Holiday* magazine in 1958, before any space probes had escaped from the Earth, and has been reproduced unchanged. All the events predicted have occurred, and many of the names I proposed for the new lunar features have in fact been adopted. (They were so inevitable that I can claim no particular credit for this!)

The Luna and Orbiter flights did, however, produce one major surprise. The statement that "there is not the slightest reason to suppose that the Moon's hidden side differs in any way from the one we can see" has turned out to be completely wrong. "Farside" is almost all mountainous, cratered highland—it has very few of the dark, low-lying "seas." No one anticipated this, and the explanation is still quite unknown.

◎

Though whole books have been written about the practical problems involved in colonizing the Moon, there is one aspect of life on our satellite which has been largely overlooked, perhaps because everyone has taken it for granted. It is an aspect which will, in fact, become important long before the first lunar landings take place, for as high-definition photographs accumulate from our rocket probes, millions of square miles of hitherto unknown territory will be dumped into the laps of geographers, scientists—and U.N. dele-

11

gates. Sometime in the 1960s, the cartographers will be faced with the biggest job of map making since exploration began.

Now when virgin territory is opened up, not only must it be mapped but its surface features must be named. This task has already been performed for the visible side of the Moon, thanks to the labors of scores of astronomers (mostly amateurs) during the last three centuries. In a way that they could scarcely have imagined, they are about to make a mark on history. For the names they gave to the lunar plains and mountains will soon pass into the vocabulary of mankind, as they blaze forth in the headlines of the future.

It is unfortunate, therefore, that so many of these names are fanciful, cumbersome, or downright inappropriate. Since all the major formations on this side of the Moon have already been labeled, it is probably too late to do much about them except in the most extreme cases. (Future lunar colonists may take violent objection to living in Hell, the Marsh of Putridity, or the Lake of Death.) The least we can do, however, is to make sure that the maps of the other side are less medieval and inconvenient.

The man who created the pattern of lunar nomenclature we are stuck with today was a Jesuit astronomer, Giovanni Riccioli, of Bologna, Italy, who published his map of the Moon in 1651. This was some forty years after Galileo had made his first telescope and astonished the world with the news that the Moon was not, as Aristotle had taught, a perfectly smooth sphere, but was even more mountainous than the Earth.

Father Riccioli's scheme for naming the new world that had been revealed in his lifetime was a consistent one, based on the fact that there are three main types of lunar formation—the dark, almost level plains, the mountain ranges, and the craters. The plains are easily visible to the naked eye, and their patterns have given rise to countless myths and legends, such as that of the angry warrior mentioned in *Hiawatha* who:

> Seized his grandmother, and threw her
> Up into the sky at midnight;

> Right against the moon he threw her;
> 'Tis her body that you see there.

In a low-powered telescope, the dark regions look very much like areas of water, and they are also at a considerably lower elevation than the brighter parts of the Moon. Though Riccioli knew perfectly well that they were dry plains, he christened them seas (*mare,* plural *maria*), oceans, lakes, bays, and so on. In the actual naming he really let his imagination go, being strongly influenced by astrological ideas and the notion that the Moon's first quarter promotes good weather while its last quarter brings storms and rain. Here are some of the more picturesque names which survive to this day on all maps of the Moon: Ocean of Storms (Oceanus Procellarum); Sea of Tranquility (Mare Tranquillitatis); Sea of Nectar (Mare Nectaris); Sea of Crises (Mare Crisium); Sea of Spring (Mare Veris); Sea of Rains (Mare Imbrium); Sea of Clouds (Mare Nubium); Bay of Rainbows (Sinus Iridum); Marsh of Sleep (Palus Somni). We can be slightly thankful that somewhere in the last three centuries Riccioli's Bay of Epidemics and Peninsula of Delirium have dropped by the wayside.

Skirting many of these dark areas are magnificent mountain ranges, some of them as high as the Himalayas, and here Riccioli took the easy way out. Following the suggestion of the astronomer Hevelius, he simply transposed terrestrial names to the Moon. So today we have the lunar Alps, Apennines, Urals, Carpathians, and Pyrenees.

The problem of finding names for the Moon's relatively few seas, lakes, bays, and mountain ranges is as nothing to that of identifying the innumerable craters. The largest map so far produced—a 300-inch-diameter chart made by the British observer H. P. Wilkins—shows about ninety thousand craters, ranging from walled plains big enough to enclose Vermont or Maryland down to tiny pits a fraction of a mile across.

Even the first crude telescopes could show at least a thousand craters, but Riccioli did not attempt to name them all. He contented himself with about two hundred, which was quite enough to start with, and the names he chose were those of great astronomers, philosophers, or scientists. The precedent thus established has, with very few exceptions, lasted to this day.

It is amusing to note how Father Riccioli's personal predilections colored his map making. An extraordinarily large number of craters bear the names of fellow Jesuits, but it is only fair to point out that they were mostly men of scientific distinction. (Even today, any large gathering of astronomers will contain a substantial number of Jesuits; the order has practically monopolized certain branches of geophysics.) When Riccioli published his map, the debate as to whether the Earth was the center of the Universe or merely another planet circling the Sun was still in full swing. Galileo had been hauled up before the Inquisition only eighteen years earlier and forced to recant his belief in a moving Earth, and Copernicus' great book, *The Revolution of the Celestial Orbs,* which founded modern astronomy, was still on the *Index Expurgatorius,* where it remained until well into the nineteenth century.

Though Riccioli could hardly ignore Galileo, the most outstanding scientist of his age, he attached his name to a small and insignificant crater tucked away near the western edge of the Moon. The conspicuous craters he reserved for the orthodox, party-line astronomers, with the result that some of the mightiest formations on the Moon are now named after long-forgotten philosophers and theologians.

Father Riccioli did, however, make a few concessions which he must have found difficult to reconcile with his conscience. Though as a faithful son of the Church he believed that the Copernican doctrine of the moving Earth was a heresy, his personal admiration for Copernicus was so great that he gave him what is perhaps the most splendid, though not the largest, crater on the face of the Moon. The most conspicuous one of all—easily visible to the naked eye—he gave to Tycho Brahe, the last great astronomer to cling to the outmoded, Earth-centered model of the Universe.

In the three centuries since Riccioli, generations of later selenographers have followed his system and given personal names to craters. The result is that the Moon has become, in Descartes's phrase, "a graveyard of astronomers." The term "graveyard" is not altogether accurate, for there are some sixty individuals alive today who have lunar craters named after them. At the last count, thirteen were Americans, the majority of the remainder British and Spanish. There are also French, Italian, Japanese, German, and Finnish representatives on the Moon, but—curiously enough—not a single living Russian, and only three dead ones. (I suspect that contemporary Soviet moon maps may show a different state of affairs.)

The right to christen a crater goes only to someone who has made a serious contribution to lunar studies, and even then the name has to be approved by the International Astronomical Union to make it official. At the moment slightly more than seven hundred lunar formations have personal names attached to them, and a study of the list is a fascinating occupation that not only produces some surprises but may also give useful hints for the future.

All together, more than thirty craters possess American names; the most celebrated is undoubtedly Benjamin Franklin, who owns a small crater (well, small for the Moon, since it's only thirty-four miles across) not far from the Sea of Serenity. And it must be admitted (*Pravda* please copy) that two United States citizens purchased their lunar immortality with hard cash, not with the imponderable currency of scientific knowledge. Yet, considering the services they rendered to astronomy, it is not likely that many will grudge the financiers Lick and Yerkes their place on the Moon.

Let us wander through the directory of lunar craters and stop at interesting or familiar names. The very first one listed is an old friend from English literature—Abenezra, or "Rabbi ben Ezra" of Browning's poem. What's *he* doing on the Moon? Well, he was a distinguished Jewish astronomer of the twelfth century and so has a perfect right to his position.

One cannot really say the same for Alexander the Great, who was put on the Moon merely to keep company with Julius Caesar.

Julius, however, has a good claim, having earned it by his reform of the calendar. And while we are on the subject of military men, it is somewhat startling to find Field Marshal Graf von Moltke owning a tiny crater, rather inappropriately close to the Sea of Tranquility. Moltke's place on the Moon was given to him (by a German astronomer, needless to say) in recognition of the fact that he persuaded the Prussian government to print an important lunar map. There is no reason to suppose that this was inspired by any early ideas of interplanetary imperialism; Moltke was himself an energetic explorer and map maker who surveyed remote parts of Asia which no European had ever visited.

Famous explorers are well represented on the Moon; among those of the past are Colombo (Columbus), Cook, Marco Polo, Pytheas, Magelhaens (Magellan), and Vasco da Gama. Coming up to more modern times, Nansen, Shackleton, Peary, Amundsen, and Scott may be found clustering round the lunar poles.

Scattered across the face of the Moon will be found the names of some of history's supreme intellects. Here is a brief listing: Archimedes, Aristotle, Darwin, Descartes, Leonardo, Einstein, Euclid, Kant, Kepler, Leibnitz, Newton, Plato, Pythagoras. Unfortunately but inevitably, the later scientists and philosophers have had a raw deal, being fobbed off with very second-rate formations. The sad case of Einstein is a good example; he has been given a sorry little crater less than thirty miles across, so near the edge of the Moon that it is almost impossible to see and might just as well be on the other side.

In contrast, the names attached to many fine craters are so obscure that only devoted historical research can uncover their origins. Others look fairly straightforward, but are quite misleading. The crater Hell, for example, was not named because of any supposed satanic associations; it commemorates Father Maximilian Hell, S.J., once Director of Vienna Observatory. Luther is not Martin, but a much later German—a nineteenth-century astronomer. The crater Pallas is not named after the Greek goddess (who already claims a minor planet) but a German explorer. Beer, disappointingly, turns out to be a Berlin banker celebrated for his astronomical studies but much less well known to the world at large than his brother, the composer Meyerbeer. And though one of the Americans enshrined on the Moon is Holden, he got there via Lick Observatory, not Hollywood. There are as yet no film stars on the Moon, though probably this is only a matter of time.

Many of the people with lunar holdings had highly checkered careers on Earth, and not a few met violent ends. Several (Lavoisier, the great chemist; Condorcet, the philosopher; Bailly, astronomer and mayor of Paris) made their exits with the aid of that highly scientific device, the guillotine. One—Cichus—was burned at the stake for necromancy, in the days when astronomy and astrology were still confused even by the intelligent.

This confusion brought disaster to the tenant of a small crater on the extreme eastern edge of the Moon. Ulug-Beg, grandson of Tamerlane, was a great patron of the sciences, and founded a splendid observatory near his capital, Samarkand. Unfortunately, when he took the natural precaution of casting the horoscope of his eldest son, he was perturbed to find that the boy was destined to kill him. Unlike most Oriental potentates, who knew how to deal with this standard situation, Ulug-Beg did not beat the young man to the draw but merely exiled him. Needless to say, he returned at the head of an invading army and, like a dutiful son, fulfilled his father's prediction. Thereafter, the historians record with a fine sense of restraint, "astronomy was no longer cultivated in Samarkand."

Another obscure name, near the south pole of the Moon, is associated with my favorite story of scientific hard luck. In the days when a journey to the Far East was a major undertaking, the French astronomer Legentil sailed to India to observe the transit of Venus across the Sun. It took place on June 6, 1761, but Legentil couldn't make the appointment; he had been delayed on the high seas by the current Anglo-French war, and when he arrived at Pondichéry the show was over. However, another transit was due in almost exactly eight years, so the stubborn astronomer decided to sit it out.

And so, in 1769, he was in the right place at the right time—but, *hélas,* the transit was completely obscured by clouds. Legentil couldn't see a thing; this, however, was not the end of his bad luck. As the next performance was not due for a hundred and five years, he packed his things and sadly sailed for France. And when he got there, he discovered that all his property had been sold, his family having assumed that by this time he must be dead.

That is enough for this side of the Moon; though one could spend a lifetime exploring it—as many have—the other hemisphere is beckoning. Yet before we cross to it, it may be well to mention briefly why there is an "other side" which we have never been able to observe. The facts are simple, but it is astonishing how poorly they are understood. One sign of the popular confusion is the expression "the dark side of the Moon." There is no such place; the Moon turns under the Sun in twenty-nine and a half days, and each face is equally illuminated during this period. Any darkness is purely temporary, as on Earth; the interchange of night and day is merely more leisurely.

Earth and Moon perform a kind of celestial dance together, and in most dances you cannot see the back of your partner's head. But imagine that the male partner, in addition to performing the dance movement, is also spinning round and round, as in some of the more energetic ballets. You then have an accurate analogy of the present Earth-Moon situation. The female partner—the Moon— sees each side of the male partner—Earth. But Earth sees only the face of the Moon, not the back of her head.

You will not be surprised to hear that this is a temporary state of affairs and that the Earth will be unable to keep it up forever. The performance is too exhausting, and in a few billion years the dance will have settled down to a sedate and stately waltz, each partner content to stare perpetually into the other's face. When that time comes, one side of the Earth will never see the Moon, as today one side of the Moon never sees the Earth.

There is not the slightest reason to suppose that the Moon's hidden side differs in any way from the one we can see. In fact, we can observe a small portion of it, because the Moon rocks slightly on her axis during the course of her revolution round the Earth, and this enables us to peer a little way over the edge. This border region is so badly foreshortened that it cannot be accurately mapped, but because of its existence we can see 60 per cent of the Moon, not merely 50 per cent.

We must assume that, as soon as we can observe the far side of the Moon, we will be confronted with some scores of mountain ranges and "seas," and at least a hundred thousand craters—all totally anonymous, all waiting to be named.

As far as the still-to-be-discovered mountain ranges are concerned, there is no problem. Earth's greatest peaks were unknown when the Moon was first mapped; there are no lunar Himalayas, Rockies, or Andes. These evocative names are crying out for mountains to match them, and we can be sure that they will be forthcoming. Also available as lunar candidates are the Appalachians, the Sierras, the Pamirs, and dozens of individual peaks such as Everest, Kilimanjaro, Whitney, Popocatepetl, Kanchenjunga, Nanda Devi . . .

The new plains—the dusky, and possibly dusty, lunar lowlands —pose some difficulties. Shall we continue to name them after bodies of water? There seems no harm in continuing the custom; it is not likely that anyone will ever be misled by it and pack skin-diving equipment on a trip to the Moon. But if the practice is continued, then the astrological and occult associations will certainly be discarded, though we need not abandon the poetic touch, which gives such charm to so many lunar place names. It may be simplest to transpose terrestrial lakes and sea; the supply is certainly adequate, and when we consider how the Moon controls the tides, the idea of lending it our oceans seems highly appropriate.

It is when we come to the craters that matters start getting complicated. Finding a hundred thousand names in a hurry would be no easy task, though luckily the problem is not quite so bad as that. Once a few hundred major formations have been named, the smaller ones can be referred to—as postal districts are in a large city—by adding letters or numbers as suffixes. This has long been standard procedure for the visible face of the Moon; thus a small crater inside the ninety-mile-diameter walled plain of Ptolemaeus might be referred to as Ptolemaeus B, or Ptolemaeus 123. (In this single case, incidentally, there are over three hundred subcraters!)

Through sheer inertia, if for no other reason, we will probably continue to give the lunar craters personal names. But whose names? The practice of honoring great scientists and philosophers is obviously worth continuing, and we might start by redressing some of the present injustices. Galileo, Newton, and Einstein should be relocated in the most splendid of the far-side craters, and their current substandard residences handed over to less important people. And Maxwell, Hertz, Roentgen, Becquerel, Curie, Rutherford, Planck, and the other makers of modern science should also be suitably rewarded.

The men who paved the way to the actual conquest of space— the great pioneers of astronautics, such as Tsiolkovsky, Oberth, and Goddard—most certainly deserve conspicuous lunar landmarks, and though there have so far been no nonhuman names on the Moon, surely a modest crater can be dedicated to Laika, the first space traveler.

It would not be difficult to find sufficient scientists, living or dead, to label the major features on both sides of the Moon. However, now that the matter is no longer of interest only to a handful of specialists, there will be claims from other quarters. Some of these will be well grounded; it is a slight scandal that there are no artists, composers, or poets on the Moon, despite all the attention they have paid to our satellite. (One exception: Leonardo has a small crater in the Moon's western—i.e., first—quadrant, but he is there because of his scientific interests, not his artistic attainments. And though there is a Wagner tucked away in the Carpathian Mountains, he turns out to be a nineteenth-century German *physiologist!*) Surely Dante, Homer, Michelangelo, Bach, Shakespeare, Milton, Goethe, Beethoven, to mention only the first who come to mind, will not be blacklisted if their names are proposed.

A slightly more controversial suggestion would be the names of the great religious leaders and reformers who have shaped the thoughts and lives not of mere millions, but of billions. Moses, Akhenaton, Asoka, Mohammed, Lao-tse, Confucius, and Gautama certainly merit apotheosis. The last three would probably have reached the Moon centuries ago, had it not been for the unexplained failure of the Chinese to invent the telescope.

The real trouble will start when the politicians and statesmen try to climb aboard the lunar bandwagon. The few already there got in by the back door and are in any event sufficiently remote not to arouse prejudices. No one today objects violently to Alexander or Caesar, and there would probably be few protests against the nominations of Washington, Napoleon, or Lincoln. But as we approach our own time, universal agreement would become more difficult. Though millions would approve of Lenin, Roosevelt, or Churchill, millions more would take a dim view of granting them lunar franchises.

The obvious solution is to allow no one on the Moon until he has been dead for a safe period—say, fifty years. That is long enough, in most cases, for greatness to be established, and for contemporary passions to evaporate. It would also eliminate the celebrities whose fame looms large in their own generation but are unknown to posterity.

If this rule is followed, then the Moon can indeed become a Roll of Honor for all mankind. Let us hope that the cartographers and photo-reconnaissance experts who must now undertake the task of naming a world do so in the spirit of responsibility and dignity it demands. We do not want to wake up one morning to find the job has been done in top secrecy by a Pentagon general who happens to be a baseball fan, or an unimaginative bureaucrat who has stuck pins at random into the Vladivostok telephone directory.

For the names we are about to write upon those unknown plains and peaks and craters will be more than chapter headings in the history of the future. They will be the words many of our grandchildren will utter when they speak of home.

Meteors

3

If you go out of doors on a clear, moonless night and look up at the sky, you will seldom have to wait for more than a few minutes before you see a meteor slide through the stars. These faint streaks of light, vanishing almost as soon as they are born, were a complete mystery to mankind for thousands of years. Until quite recent times, indeed, it was not even realized that they had any connection with the other heavenly bodies; they were considered to be purely atmospheric phenomena, perhaps something akin to lightning. The very word "meteor," with its obvious kinship to "meteorology," is a survival of this old belief.

Ours is an age in which subjects that once seemed of no interest to anyone except a few ivory-tower scientists have suddenly become of overwhelming practical (and, alas, all too often military) importance. So it is with the transient lines of fire in the night sky. Within the last few years, the study of meteors has become the concern of research teams all over the world, and tomorrow it may determine the very survival of great nations.

The fact that meteor trails are caused by fragments of matter from outer space entering the Earth's atmosphere at enormous velocities is now known to almost everybody. Yet it was not until the beginning of the last century that astronomers accepted this fact, and then only after a determined rear-guard action. Science (if there is such a thing as Science with a capital S!) is often accused of being orthodox and unwilling to give heed to new ideas, and there are times when the charge has some truth in it. The argument over the origin of meteors is a perfect example of this.

Though there had been reports in all times and from all lands of stones falling from the sky, the scientists of the French Academy, in the closing years of the eighteenth century—when it was confidently believed that the Age of Reason had dawned—dismissed all such tales as superstitious nonsense. They reacted, in fact, much as an astronomer of today would react when confronted with a typical flying-saucer report—though it by no means follows that the sequel will be similar. And then, in 1803, almost as if Nature had determined to teach the skeptical scientists a lesson, a great shower of meteoric stones fell in Normandy—geographically speaking, on the Academy's doorstep. Thereafter no one doubted the fact that objects from outer space entered the Earth's atmosphere and occasionally reached the surface.

It was another thirty years before meteors attracted much further attention; then they did so with a spectacle the like of which has seldom been matched before or since. Listen to the words of a South Carolina planter, describing what happened on the night of November 11, 1833:

I was suddenly awakened by the most distressing cries that ever fell on my ears. Shrieks of horror and cries of mercy I could hear from most of the Negroes of the three plantations. . . . While earnestly listening for the cause I heard a faint voice near the door, calling my name. I arose and, taking my sword, stood at the door. At this moment I heard the same voice still beseeching me to rise and saying, "Oh my God, the world is on fire!" I then opened the door, and it is difficult to say which excited me most—the awfulness of the scene, or the distressing cries of the Negroes. Upwards of a hundred lay prostrate on the ground, some speechless, and some with the bitterest cries, but with their hands raised, imploring God to save the world and them. The scene was truly awful: for never did rain fall much thicker than the meteors fell towards the Earth: East, West, North and South it was the same.

Such was the great shower of 1833, which dramatically demon-strated that meteors could occur not only as sporadic wanderers but also in enormous clusters or streams. As a result of many years of observation, large numbers of these meteor showers have been iden-tified and their dates of arrival noted. For example, around the twelfth of August every year, meteors will be seen streaking from the heart of the constellation Perseus at the rate of about one a minute. And between the fourteenth and sixteenth of November, in the constellation of Leo, the shower which caused such alarm over the southern states in 1833 still puts on an annual display—though in most years it is so feeble that one would never notice it unless one was on the lookout.

Until the close of the Second World War, the study of meteors was a somewhat neglected branch of astronomy. Since they are so transient and unpredictable, they cannot be watched through tele-scopes except by pure chance, and hence almost all observations until recently were naked-eye ones made by amateur astronomers with no equipment but a notebook, a watch, a thorough knowledge of the constellations, unlimited patience, and a complete indiffer-ence to cold and fatigue. These devoted souls would spend their nights watching the stars, and every time a meteor flashed across the sky would note its duration and would pinpoint the beginning and end of its track. It may seem surprising to those who think that astronomers have to work with huge and expensive instru-ments that anything useful could be discovered by such simple means. Yet it was from thousands of such naked-eye observations that almost all our knowledge of meteors was derived, until the in-vention of radar gave astronomy a new and unexpected tool of tre-mendous power.

Behind this there is a story of war and science that is still largely untold. During the late thirties, Britain began to build the chain of radar stations without which the Royal Air Force could never have held the Luftwaffe at bay. The men who designed and constructed the three-hundred-foot-high towers along the east coast of England changed the history of the world by defeating Goering's bombers in the Battle of Britain. Three years later, in 1944, they were called upon again to fight the weapon that made those bombers obsolete.

The V-2 rockets which the radar chain now had to detect traveled ten times as fast as any bomber, and twenty times as high. Nevertheless, the hastily modified radar picked them up. It also picked up something else—something that produced strange echoes seventy or eighty miles above the earth.

These echoes, it was soon discovered, were due to meteors—or, to be more accurate, to the trails of intensely heated gas which meteors leave in their wake as they plunge into the upper atmosphere at speeds often exceeding a hundred thousand miles an hour. It was obviously a matter of great importance to distinguish between the echoes caused by meteors and those produced by rockets. And it is even more important now that those rockets can carry a million tons of explosive power instead of the miserable one ton of the quaint, old-fashioned V-2.

After the war, when radar apparatus was available for more peaceful uses, regular watch was kept for meteors at "radio observatories" throughout the world. The enormous advantage of radar for this work lies in the fact that it is independent of weather conditions and can operate just as well during daylight as at night. Previously, there had been no way of observing meteors except after dark—and even then only when there was no Moon to flood the sky with light.

It is hardly surprising, therefore, that some remarkable discoveries were very quickly made. The most spectacular of these was undoubtedly the detection, by the group of radio astronomers at Manchester, England, of great meteor showers that occur during the hours of daylight and so are quite invisible to the eye. Every summer, showers take place which if they occurred after nightfall would produce a display almost as dramatic as the one of 1833. Between June and August, vast belts of meteors are sweeping unseen, and until today unsuspected, across the daylight skies of Earth.

Continuous watches are now being kept by automatic equipment which, as soon as a meteor is located, photographs its radar echo on a cathode-ray tube. From this it is possible to calculate the meteor's height and velocity, and thus the orbit it was following through space before it met its doom. This radar watch has already settled one question that astronomers had been fighting about furiously for more than a generation.

One school of thought maintained that a substantial proportion of meteors did not belong to the Solar System at all but came from interstellar space—that there were, in other words, vast streams of meteoric matter flowing between the stars. The evidence for this startling theory was quite strong—indeed, at first sight overwhelming. When the velocities of meteors were measured by the indirect methods which were the only ones available before radar, many were found to be traveling so fast that they could not possibly be revolving around the Sun. In the Earth's neighborhood, any object moving at more than 94,000 miles an hour could only be a visitor to the Solar System, not a permanent resident. This is the limiting speed above which the Sun can no longer keep a body under its gravitational control. Anything moving faster than this, accordingly, must have fallen into the Solar System from outside and would shoot out of it again after doing a tight turn round the Sun.

The more accurate radar methods proved conclusively that meteors traveling faster than this solar speed limit did not exist; all meteors, therefore, are as much captives of the Sun as are Earth and the other planets and revolve around it in similar closed orbits.

Although meteors do not travel faster than 94,000 miles an hour with respect to the Sun in our part of the Solar System, the velocities with which they hit our atmosphere can be far higher than this, since the Earth itself is racing along its orbit at 66,000 miles an hour. When Earth and meteor hit head on, therefore, their combined speed may be as much as 160,000 miles an hour—a velocity that would take one to the Moon in ninety minutes.

On the other hand, when a meteor catches up to the Earth from the rear its speed of approach is relatively low, and this sometimes produces a remarkable effect. Though most meteor trails flash out and vanish in a second, when one of these "slow" meteors enters the atmosphere it may make a sedate and dignified—not to say impressive—progress across the sky. There have even been occasions when an entire procession of meteors has put on such a display, apparently for the express purpose of adding to the flying-saucer mythos. (I'm sorry to raise that subject again, but it's never far away where meteors are concerned.)

It is very important to draw a clear distinction between meteors themselves and the trails they produce in the sky when they happen to hit the Earth's atmosphere. It is these trails that are observed both by the eye and by the electronic senses of the radar telescope; the meteors are far too small to be detected. There is a close analogy here with something we have all witnessed when a jet plane passes high overhead. Often the vapor trail can be seen stretching for miles across the deep blue of the stratosphere, but of the plane itself there is no sign.

In the case of meteors, the disparity between the size of the trail and the object causing it is far more extreme. Even a very bright meteor—one producing a burst of light outshining all the stars put together—is only about half an inch in diameter. Such a giant is quite rare; perhaps a thousand hit the entire Earth every hour. Anyone who considers that this makes them hardly uncommon should remember that the Earth is a rather large object, and that in an hour it carves out a tunnel through space 8,000 miles in diameter and 66,000 miles long.

The total number of meteors, of *all* sizes, that hit the Earth every hour is enormous—probably in the billions. But the vast majority of these are smaller than grains of sand; most, indeed, are specks of dust that would be invisible to the eye.

Ever since space travel and artificial satellites began to be considered seriously, a good deal of attention has been paid to the possible hazard that meteors might represent. As long ago as 1946 the Rand Corporation concerned itself with this problem on behalf of the Air Force, and made public its findings in an unclassified report. The results were reassuring and have since been confirmed by satellite observations; meteors are very much less of a danger to space travelers than automobiles are to practically everybody. You would die of old age on an interplanetary journey before you met a meteor large enough to do any serious damage, though it is possible that there may be enough meteoric dust around in space to "sand-blast" windows and optical surfaces after a few years of continuous operations. Meteors may be a nuisance, but they will certainly not be a menace.

About ten times a day the Earth encounters a meteor that is sufficiently large not to be consumed by the friction of its passage through the atmosphere and manages to reach the surface intact. It is then termed a meteorite, passing from the jurisdiction of astronomy to that of meteoritics (studied, heaven help them, by meteoriticists. Try to say *that* quickly after the fourth or fifth martini). Since these falling bodies are the only samples we have of matter from outside the Earth, they are of great interest to science, and nowadays any report of a falling meteorite sparks off something like a gold rush.

The average meteorite is an unprepossessing lump of stone or nickel-iron which looks as if it had been picked up from a slag heap. In a sense, indeed, it is a lump of cosmic slag—possibly part of the debris left over when the planets were formed, at least five billion years ago. Once or twice every century, really large meteorites hit the Earth; it happened in Siberia in 1908 and again in 1947. Several hundred tons of iron and stone plunging down through the atmosphere at ten or more times the speed of an artillery shell can produce a blast wave greater than that of an atomic bomb; the 1908 meteorite felled a forest, snapping off tree trunks for miles around so that they lay like matchsticks pointing away from the impact area.

During the history of the Earth there must have been thousands of such collisions, but the effects of weather and vegetation have obliterated the evidence—and it should also be remembered that most meteorites must come down in the sea. Until recently, the famous Meteor Crater in Arizona was the largest known imprint of one of these prehistoric catastrophes; with a diameter of over four thousand feet, it is very impressive, especially from the air.

During the Second World War, United States and Canadian Air Force pilots noticed a curious circular lake in the frozen wastes of Northern Quebec, and this has now been found to mark the site of a meteor crater more than eleven thousand feet in diameter. The Ungava Crater, as it has been named, has certainly been there for many thousands of years, for the glaciers of the last Ice Age have ground their way across it and retreated again, leaving unmistakable marks of their passage. So, though the Ungava Crater is more than twice as large as its Arizona rival, it is not in the same pristine condition, and much of the evidence of its formation has probably been erased.

There is little doubt that air surveys will reveal many more formations of this type, some of them in populated areas. At the village of Cabrerolles in Southern France, for instance, lies a group of craters which no one had ever noticed because they had been completely overgrown with vegetation. One of them is occupied by a vineyard. It is not yet certain that they were caused by meteorites, and anyone who knows much about the French peasantry will realize that the scientists may have to do some hard bargaining before they can start digging for nickel-iron fragments.

It is always possible that a large meteorite may fall on a city— and one can guess the consequences if, by doubly bad luck, this should happen during a period of international tension. In the whole of recorded history, however, there are less than half a dozen cases of deaths from falling meteorites, and a recent statistical analysis showed that there is only about one chance in three that a single member of the human race will be hit by a meteorite during the entire twentieth century. An insurance company wishing to make the headlines, therefore, would not be taking much of a risk if it offered $10 billion compensation to any client meeting this unusual mishap. If the phrase "almost unique" can be justified in strict logic, here is a case for employing it.

Yet, though the chance of a personal encounter with a meteorite is remote, these visitors from outer space now affect the lives of every one of us. Today, the problem that first confronted the British radar experts during the closing months of the war has become one of vital importance. How is one to distinguish between an intercontinental ballistic missile and a meteor that may be traveling at the same speed and at the same height? A few minutes' wait will give the answer, of course; but then it may be a little too late.

There is considerable evidence that without meteors we would have no long-range radio communication. The only way that radio waves can get round the curve of the Earth is by bouncing off the ionized layers in the upper atmosphere some seventy miles above our heads. Why the air in this region should act as a kind of radio mirror is still something of a mystery; during the daytime, it is true, the Sun's rays are able to keep it electrically charged, but that does not explain how it persists at night. It is now fairly certain that the continuous gentle rain of meteor dust from space is responsible for at least one of the electrified layers which enable us to send our voices round the world.

Some recent research, started in Australia, has shown that meteors may, after all, have some association with meteorology. The link is an unexpected one, but if it can be established it will be of very great practical importance. It appears that our small-scale attempts to produce rain by "seeding" clouds with dry ice and other substances have been anticipated by Nature; the ceaseless shower of meteoric dust filtering down from the stars may have the same effect. Long-range weather prediction, therefore, will have to take account of the meteor streams which the Earth encounters in its passage through space.

It would be hard to find a better example of the way in which apparently unrelated branches of science prove to be closely connected. Though the laws which govern the Universe may be simple, the effects which they produce can be exceedingly complex. One of the giant planets may deflect a meteor stream half a billion miles from Earth, so that ages later our world encounters an abnormally high concentration of dust as it sweeps along its orbit. And so an event far off in space and time can cause rains and floods which may destroy many lives and undo the work of generations of men.

A hundred years ago the greatest poet of the Victorian Age wrote these words:

> Now sleeps the crimson petal, now the white;
> Nor waves the cypress in the palace walk;
> Nor winks the gold fin in the porphyry font:
> The fire-fly wakens: waken thou with me.
>
> Now droops the milk-white peacock like a ghost,
> And like a ghost she glimmers on to me.
>
> Now lies the Earth all Danaë to the stars,
> And all thy heart lies open unto me.
> Now slides the silent meteor on, and leaves
> A shining furrow, as thy thoughts in me.

A different description, perhaps, from the one that science gives, and perhaps some may prefer it. Yet both are equally valid, and why should we not appreciate the beauty of that "shining furrow" all the more, now that we are beginning to uncover its secrets?

Postscript

The preceding article contains a good example of the danger of betting on anything—however "certain." Despite the incredible odds, a human being *has* been hit by a meteorite. In December 1954, shortly before this essay was written, a Mrs. Hewlitt Hodges of Sylacauga, Alabama, was grazed by a ten-pound meteorite that came through the roof of her house. The first case of a meteorite hitting an automobile also occurred in the United States—at Benld, Illinois, in September 1938.

It is now apparent that meteor craters are much more common on Earth than had been supposed; the very largest were simply too big to be seen before aerial photography became common, and there may be some that will be detected only from space. One of the largest of these "astroblemes" (literally, "star wounds") yet discovered is the huge Vreedefort Structure, over *thirty miles* in diameter, in South Africa.

The Star of the Magi

This article was written for the December 1954 issue of *Holiday* magazine, but I have not changed the opening paragraph, because almost *every* Christmas Venus is a brilliant object either in the morning or the evening sky.

Readers of my fiction will recognize in this essay the origins of the short story "The Star."

Where is he that is born King of the Jews? for we have seen his star in the east, and are come to worship him.

Go out of doors any morning this December and look up at the eastern sky an hour or so before dawn. You will see there one of the most beautiful sights in all the heavens—a blazing, blue-white beacon, many times brighter than Sirius, the most brilliant of the stars. Apart from the Moon itself, it will be the brightest object you will ever see in the night sky. It will still be visible even when the Sun rises; you will even be able to find it at midday if you know exactly where to look.

It is the planet Venus, our sister world, reflecting across the gulfs of space the sunlight glancing from her unbroken cloud shield. Every nineteen months she appears in the morning sky, rising shortly before the Sun, and all who see this brilliant herald of the

Christmas dawn will inevitably be reminded of the star that led the Magi to Bethlehem.

What was that star, assuming that it had some natural explanation? Could it, in fact, have been Venus? At least one book has been written to prove this theory, but it will not stand up to serious examination. To all the people of the Eastern world, Venus was one of the most familiar objects in the sky. Even today, she serves as a kind of alarm clock to the Arab nomads. When she rises, it is time to start moving, to make as much progress as possible before the Sun begins to blast the desert with its heat. For thousands of years, shining more brilliantly than we ever see her in our cloudy northern skies, she has watched the camps struck and the caravans begin to move.

Even to the ordinary, uneducated Jews of Herod's kingdom, there could have been nothing in the least remarkable about Venus. And the Magi were no ordinary men; they were certainly experts on astronomy, and must have known the movements of the planets better than do ninety-nine people out of a hundred today. To explain the Star of Bethlehem we must look elsewhere.

The Bible gives us very few clues; all that we can do is to consider some possibilities which at this distance in time can be neither proved nor disproved. One of these possibilities—the most spectacular and awe-inspiring of all—has been discovered only in the last few years, but let us first look at some of the earlier theories.

In addition to Venus, there are four other planets visible to the naked eye—Mercury, Mars, Jupiter, and Saturn. During their movements across the sky, two planets may sometimes appear to pass very close to one another—though in reality, of course, they are actually millions of miles apart.

Such occurrences are called "conjunctions"; on occasion they may be so close that the planets cannot be separated by the naked eye. This happened for Mars and Venus on October 4, 1953, when for a short while the two planets appeared to be fused together to give a single star. Such a spectacle is rare enough to be very striking, and the great astronomer Johannes Kepler devoted much time to proving that the Star of Bethlehem was a special conjunction of Jupiter and Saturn. The planets passed very close together (once again, remember, this was purely from the Earth's point of view— in reality they were half a billion miles apart!) in May, 7 B.C. This is quite near the date of Christ's birth, which probably took place in the spring of 7 or 6 B.C. (This still surprises most people, but as Herod is known to have died early in 4 B.C., Christ must have been born before 5 B.C. We should add six years to the calendar for A.D. to mean what it says.)

Kepler's proposal, however, is as unconvincing as the Venus theory. Better calculations than those he was able to make in the seventeenth century have shown that this particular conjunction was not a very close one, and the planets were always far enough apart to be easily separated by the eye. Moreover, there was a closer conjunction in 66 B.C., which on Kepler's theory should have brought a delegation of wise men to Bethlehem sixty years too soon!

In any case, the Magi could be expected to be as familiar with such events as with all other planetary movements, and the Biblical account also indicates that the Star of Bethlehem was visible over a period of weeks (it must have taken the Magi a considerable time to reach Judea, have their interview with Herod, and then go on to Bethlehem). The conjunction of two planets lasts only a very few days, since they soon separate in the sky and go once more upon their individual ways.

We can get over the difficulty if we assume that the Magi were astrologers ("Magi" and "magician" have a common root) and had somehow deduced the birth of the Messiah from a particular configuration of the planets, which to them, if to no one else, had a unique significance. It is an interesting fact that the Jupiter-Saturn conjunction of 7 B.C. occurred in the constellation Pisces, the Fish. Now though the ancient Jews were too sensible to believe in astrology, the constellation Pisces was supposed to be connected with them. Anything peculiar happening in Pisces would, naturally, direct the attention of Oriental astrologers toward Jerusalem.

This theory is simple and plausible, but a little disappointing.

One would like to think that the Star of Bethlehem was something more dramatic and not anything to do with the familiar planets whose behavior had been perfectly well known for thousands of years before the birth of Christ. Of course, if one accepts as *literally* true the statement that "the star, which they saw in the east, *went before them, till it came and stood over where the young Child was,*" no natural explanation is possible. Any heavenly body—star, planet, comet, or whatever—must share in the normal movement of the sky, rising in the east and setting some hours later in the west. Only the Pole Star, because it lies on the invisible axis of the turning Earth, appears unmoving in the sky and can act as a fixed and constant guide.

But the phrase, "went before them," like so much else in the Bible, can be interpreted in many ways. It may be that the star—whatever it might have been—was so close to the Sun that it could be seen only for a short period near dawn, and so would never have been visible except in the eastern sky. Like Venus when she is a morning star, it might have risen shortly before the Sun, then been lost in the glare of the new day before it could climb very far up the sky. The wise men would thus have seen it ahead of them at the beginning of each day, and then lost it in the dawn before it had veered around to the south. Many other readings are also possible.

Very well, then, can we discover some astronomical phenomenon sufficiently startling to surprise men completely familiar with the movements of the stars and planets and which fits the Biblical text?

Let's see if a comet would answer the specification. There have been no really spectacular comets in this century—though there were several in the 1800s—and most people do not know what they look like or how they behave. They even confuse them with meteors, which any observer is bound to see if he goes out on a clear night and watches the sky for half an hour.

No two classes of object could be more different. A meteor is a speck of matter, usually smaller than a grain of sand, which burns itself up by friction as it tears through the outer layers of Earth's atmosphere. But a comet may be millions of times larger than the entire Earth, and may dominate the night sky for weeks on end. A really great comet may look like a searchlight shining across the stars, and it is not surprising that such a portentous object always caused alarm when it appeared in the heavens. As Calpurnia said to Caesar:

> When beggars die, there are no comets seen;
> The heavens themselves blaze forth the death of princes.

Most comets have a bright, starlike core, or nucleus, which is completely dwarfed by their enormous tail—a luminous appendage which may be in the shape of a narrow beam or a broad, diffuse fan. At first sight it would seem very unlikely that anyone would call such an object a star, but as a matter of fact in old records comets are sometimes referred to, not inaptly, as "hairy stars."

Comets are unpredictable: the great ones appear without warning, come racing in through the planets, bank sharply around the Sun, and then head out toward the stars, not to be seen again for hundreds or even millions of years. Only a few large comets—such as Halley's—have relatively short periods and have been observed on many occasions. Halley's comet, which takes seventy-five years to go around its orbit, has managed to put in an appearance at several historic events. It was visible just before the sack of Jerusalem in A.D. 66, and before the Norman invasion of England in A.D. 1066. Of course, in ancient times (or modern ones, for that matter) it was never very difficult to find a suitable disaster to attribute to any given comet. It is not surprising, therefore, that their reputation as portents of evil lasted for so long.

It is perfectly possible that a comet appeared just before the birth of Christ. Attempts have been made, without success, to see if any of the known comets were visible around that date. (Halley's, as will be seen from the figures above, was just a few years too early on its last appearance before the fall of Jerusalem.) But the number of comets whose paths and periods we do know is very small compared with the colossal number that undoubtedly exists.

If a comet did shine over Bethlehem, it may not be seen again from Earth for a hundred thousand years.

We can picture it in that Oriental dawn—a band of light streaming up from the eastern horizon, perhaps stretching vertically toward the zenith. The tail of a comet always points away from the Sun; the comet would appear, therefore, like a great arrow, aimed at the east. As the Sun rose, it would fade into invisibility; but the next morning, it would be in almost the same place, still directing the travelers to their goal. It might be visible for weeks before it disappeared once more into the depths of space.

The picture is a dramatic and attractive one. It may even be the correct explanation; one day, perhaps, we shall know.

But there is yet another theory, and this is the one which most astronomers would probably accept today. It makes the other explanations look very trivial and commonplace indeed, for it leads us to contemplate one of the most astonishing—and terrifying—events yet discovered in the whole realm of nature.

We will forget now about planets and comets and the other denizens of our own tight little Solar System. Let us go out across *real* space, right out to the stars—those other suns, many far greater than our own, which sheer distance has dwarfed to dimensionless points of light.

Most of the stars shine with unwavering brilliance, century after century. Sirius appears now exactly as it did to Moses, as it did to Neanderthal man, as it did to the dinosaurs—if they ever bothered to look at the night sky. Its brilliance has changed little during the entire history of our Earth and will be the same a billion years from now.

But there are some stars—the so-called "novae," or new stars—which through internal causes suddenly become celestial atomic bombs. Such a star may explode so violently that it leaps a hundred-thousand-fold in brilliance within a few hours. One night it may be invisible to the naked eye; on the next, it may dominate the sky. If our Sun became such a nova, Earth would melt to slag and puff into vapor in a matter of minutes, and only the outermost of the planets would survive.

Novae are not uncommon; many are observed every year, though few are near enough to be visible except through telescopes. They are the routine, everyday disasters of the Universe.

Two or three times in every thousand years, however, there occurs something which makes a mere nova about as inconspicuous as a firefly at noon. When a star becomes a *super*nova, its brilliance may increase not by a hundred thousand but by a *billion* in the course of a few hours. The last time such an event was witnessed by human eyes was in A.D. 1604; there was another supernova in A.D. 1572 (so brilliant that it was visible in broad daylight); and the Chinese astronomers recorded one in A.D. 1054. It is quite possible that the Bethlehem star was such a supernova, and if so one can draw some very surprising conclusions.

We'll assume that Supernova Bethlehem was about as bright as the supernova of A.D. 1572—often called "Tycho's star," after the great astronomer who observed it at the time. Since this star could be seen by day, it must have been as brilliant as Venus. As we also know that a supernova is, in reality, at least a hundred million times more brilliant than our own Sun, a very simple calculation tells us how far away it must have been for its *apparent* brightness to equal that of Venus.

It turns out that Supernova Bethlehem was more than three thousand light years—or, if you prefer, 18 quadrillion miles—away. That means that its light had been traveling for at least three thousand years before it reached Earth and Bethlehem, so that the awesome catastrophe of which it was the symbol took place five thousand years ago, when the Great Pyramid was still fresh from the builders.

Let us, in imagination, cross the gulfs of space and time and go back to the moment of the catastrophe. We might find ourselves watching an ordinary star—a sun, perhaps, no different from our own. There may have been planets circling it; we do not know how common planets are in the scheme of the Universe, and how many suns have these small companions. But there is no reason to think that they are rare, and many novae must be the funeral pyres of worlds, and perhaps races, greater than ours.

There is no warning at all—only a steadily rising intensity of the sun's light. Within minutes the change is noticeable; within an hour, the nearer worlds are burning. The star is expanding like a balloon, blasting off shells of gas at a million miles an hour as it blows its outer layers into space. Within a day, it is shining with such supernal brilliance that it gives off more light than *all the other suns in the Universe combined*. If it had planets, they are now no more than flecks of flame in the still-expanding shells of fire. The conflagration will burn for weeks before the dying star collapses back into quiescence.

But let us consider what happens to the light of the nova, which moves a thousand times more swiftly than the blast wave of the explosion. It will spread out into space, and after four or five years it will reach the next star. If there are planets circling that star, they will suddenly be illuminated by a second sun. It will give them no appreciable heat, but will be bright enough to banish night completely, for it will be more than a thousand times more luminous than our full Moon. All that light will come from a single blazing point, since even from its nearest neighbor Supernova Bethlehem would appear too small to show a disk.

Century after century, the shell of light will continue to expand around its source. It will flash past countless suns and flare briefly in the skies of their planets. Indeed, on the most conservative estimate, this great new star must have shone over thousands of worlds before its light reached Earth—and to all those worlds it appeared far, far brighter than it did to the men it led to Judea.

For as the shell of light expanded, it faded also. Remember, by the time it reached Bethlehem it was spread over the surface of a sphere six thousand light-years across. A thousand years earlier, when Homer was singing the song of Troy, the nova would have appeared twice as brilliant to any watchers further upstream, as it were, to the time and place of the explosion.

That is a strange thought; there is a stranger one to come. For the light of Supernova Bethlehem is still flooding out through space; it has left Earth far behind in the twenty centuries that have elapsed since men saw it for the first and last time. Now that light is spread over a sphere ten thousand light-years across and must be correspondingly fainter. It is simple to calculate how bright the supernova must be to any beings who may be seeing it now as a new star in *their* skies. To them, it will still be far more brilliant than any other star in the entire heavens, for its brightness will have fallen only by 50 per cent on its extra two thousand years of travel.

At this very moment, therefore, the Star of Bethlehem may still be shining in the skies of countless worlds, circling far suns. Any watchers on those worlds will see its sudden appearance and its slow fading, just as the Magi did two thousand years ago when the expanding shell of light swept past the Earth. And for thousands of years to come, as its radiance ebbs out toward the frontiers of the Universe, Supernova Bethlehem will still have power to startle all who see it, wherever—and whatever—they may be.

Astronomy, as nothing else can do, teaches men humility. We know now that our Sun is merely one undistinguished member of a vast family of stars, and no longer think of ourselves as being at the center of creation. Yet it is strange to think that before its light fades away below the limits of vision, we may have shared the Star of Bethlehem with the beings of perhaps a million worlds—and that to many of them, nearer to the source of the explosion, it must have been a far more wonderful sight than ever it was to any eyes on earth.

What did they make of it—and did it bring them good tidings, or ill?

Postscript

Many planetariums put on a special display at Christmas in which the possible explanations of the Star of the Nativity are discussed and demonstrated. New York City's Hayden Planetarium, for example, has a particularly impressive and moving program, "The Christmas Sky," every December, which should not be missed by anyone who has an opportunity to see it.

II. *Outward from Earth*

Vacation in Vacuum

This essay was commissioned by *Holiday* magazine in 1953—four years before Sputnik 1. At the time, most readers must have thought that orbiting hotels were the wildest of fantasy, but now Barron Hilton firmly expects to be running such establishments before the year 2001 dawns.

And talking of *2001* (as we shall be) here is the original inspiration of the space-station sequence in the movie; Stanley Kubrick constructed the "Sky Grill," full-sized, at M-G-M's Borehamwood Studios.

I must confess that I now have my doubts about the practicability—and stability—of the spherical swimming pool; but a hollow cylindrical one could certainly be built, and would be just as much fun.

When the United States and the U.S.S.R. started building the first satellite stations, back in the 1960s, the idea that they would one day become health resorts and embarkation points for space-bound vacationers would have seemed slightly fantastic. Yet it was no more fantastic, of course, than the fact that since the beginning of the century the human race had deserted the sea and lifted its commerce into the air. If anyone had dared to prophesy *that* miracle when the Wright Brothers made their first nervous hop in

1903, he would have been laughed to scorn. And even fifty years later, though there were many who realized that space stations might have military and scientific uses, there were very few who looked beyond these to the day when they would become part of everyday life.

Well, perhaps that is a slight exaggeration. Even today, relatively few people have actually *been* to a space station, but there can be nobody who has not seen one with his own eyes. If you live near the equator you have a fine selection to choose from: you can see not only the outer stations but the close refueling satellites that hug the edge of the atmosphere and are so near the Earth that the curve of the planet hides them from observers in high latitudes. In the daytime they are bright stars, easily visible when the sky is clear, sweeping from horizon to horizon in a matter of minutes. And, of course, they move backward, from west to east, because they race around their tight little orbits so much more quickly than the Earth itself turns on its axis.

At night, they are the brightest stars in the sky, and you can see them move even as you watch. You'll have to look for them low down near the horizon, for as they rise up they disappear into Earth's vast, invisible shadow, winking suddenly out of existence as they go into eclipse and no longer catch the light of the Sun. Sometimes, if you are lucky, you may see a star snuffed out for a few seconds as a space station moves silently across it up there in the emptiness beyond the atmosphere. But the stations are so tiny, and the sky so vast, that you'll have to watch for many nights before you'll see this happen.

Let's go up there into the shining darkness of space, into that paradoxical world where intense heat and unimaginable cold exist together, where dawn and dusk are separated by minutes, not by hours. Yet before we begin the journey, we'll glance back for a moment into the twentieth century, to remind ourselves how so much of what we now take for granted first came into being.

It was around 1925 that scientists first became seriously interested in space stations as refueling stops for interplanetary rockets.

Back at that time, of course, there weren't any rockets, interplanetary or otherwise, and the general public never heard about the idea. It didn't hit the headlines until 1948, soon after the end of the Second World War. The United States military experts had been studying the results of German war research and had been staggered by what they found. They were now seriously investigating, the Secretary of Defense announced, the possibilities of "space platforms" for military use.

Looking at the newspapers of that time, it's amusing to note the reactions. Many editors asked sarcastically how such platforms could possibly stay up there in the sky. Apparently they'd never bothered to consider how the Moon "stayed up," and so had not realized that the proposed artificial satellites would obey exactly the same laws as the natural ones.

Slowly, during the 1950s and '60s, the idea was accepted by the general public as well as the military. As rockets reached greater speeds and altitudes, the goal of the Earth satellite vehicle came nearer to realization, until at last a few instruments were flung out into space, never to return to the atmosphere. That was the first frail rung on the ladder that would lead to the planets.

It was still many years before real man-carrying space stations, and not mere automatic missiles, were constructed from prefabricated parts ferried up by rocket and assembled in space. By the end of the twentieth century there were dozens of military reconnaissance units, meteorological stations, and astronomical observatories circling the Earth at various distances, carrying crews of up to twenty men in conditions almost as cramped as in the old-time submarines. They were the first forerunners of the spacious orbital cities we have today—the nuclei around which the later satellites were built, just as on Earth itself great capitals once grew from ancient villages or fortified camps.

The ordinary space traveler sees only the inner station—Space Station One—as he transfers from the Earth ferry rocket to the liner that's taking him to Mars or Venus. It's the nearest of all the satellites, a mere three hundred miles up—too close, therefore, to give one a really good view of Earth. If you want to see the planet as a whole, you've got to travel out to one of the more distant stations. We'll start our tour, therefore, more than ten thousand miles out, in the most luxurious of all the satellites—Sky Hotel.

Even today, with all our modern developments in rocketry, it's highly doubtful that a hotel in space would be a commercial proposition. However, Sky Hotel draws its income from many subsidiary sources; it's not merely patronized from Earth. The staffs from the other satellites take their vacations there, as it's cheaper for them to do that than to pay the fares down to Earth and up again. Moreover, Sky Hotel has pretty large shares in the relay stations, which we'll be visiting later in our trip.

The hotel is in two sections—the part with gravity and the part without. When you first see it from your approaching rocket, you'll think you're about to land on the planet Saturn. Hanging there in space ahead of you is a great ball, with a ring surrounding it but not touching it at any point. The ball is motionless, while the ring slowly revolves.

When the pilot has jockeyed the rocket over to the ball, you'll realize just how big the hotel is. Your ship will seem like a toy when it couples itself up to the mooring socket on the axis of the station and the air locks are joined together so that you can go aboard. The hotel staff will collect both you and your luggage, for most people are pretty helpless under zero gravity for the first few hours. But, believe me, it's an experience worth getting used to.

Sky Hotel has, by ingenious design, managed to get the best of both worlds. Most vacationers go up there to enjoy the fun and games under zero-gee, but weightlessness is not so amusing when you want to eat a meal or take a bath, and some people find it impossible to sleep under free-fall conditions. Hence the dual-purpose design of the hotel. The central ball contains the gymnasiums and that fantastic swimming pool we'll be visiting presently, while over in the ring are the bedrooms, lounges, and restaurant. As the ring rotates, centrifugal force gives everyone inside it a feeling of weight which can't be distinguished from the real thing. It's not so powerful, though—at the outer rim of the hotel you'll weigh only half as much as you would on Earth.

And there's one other difference between gravity home on Earth and the imitation variety in the hotel. Because "UP" always points to the center of the ring—to the invisible axle on which it turns— all the floors are curved, like the inside of a drum. If you could see right across the hotel—and maybe it's just as well that you can't— you'd see that the people on the other side were upside down, with their heads pointing toward you.

It's only in the Sky Grill—the largest room in the ring—that this effect is at all noticeable. When you're dining, your table seems to be at the bottom of a smoothly curving valley, while everyone else is sitting at improbable angles farther up the slope. The more distant they are from you, the more canted toward you they will be, until eventually it seems that they must be glued to the wall. It's a fascinating sight watching a waiter come down the slope with a trayful of beers. At first you won't be able to believe your eyes— why don't the glasses spill? Then as he approaches he'll veer over to what you—but nobody else—consider to be the vertical, and you'll breathe a sigh of relief.

Of course, there's nothing mysterious about all this. Centrifugal force can produce exactly the same effect down on Earth if you whirl a bucket at the end of a rope. But I advise you to do the experiment out of doors, and to use water rather than beer.

Most of the hotel's residents divide their time more or less equally between the gee and the zero-gee parts—between the ring and the ball, in other words. The kids are an exception. It's a job luring them away from weightlessness, even for meals, so they spend almost all their time in the ball. There is a snack bar over there, where you can get drinks served in plastic bulbs so that you can squirt the liquid straight into your mouth. That's the theory— and it works, too. But the kids usually prefer less efficient methods, and promptly empty their bulbs into the air. It's quite a sight watching a budding space cadet chasing a ball of Coke as it drifts slowly from point to point and eventually splatters messily on one of the walls.

Traveling between the stationary ball and the spinning ring that surrounds it is another of the novelties of space-station life. The trip is made in a kind of pressurized elevator cage, running around a track on the inside of the ring. It's a queer sensation, feeling your weight ebb away as you move across to the ball and centrifugal force vanishes.

The hotel is full of ingenious mechanisms and gadgets like this. Most of them you'll take for granted and may never even see unless you get one of the engineers to take you behind the scenes. Then you may be shown around the air purifiers that crack the carbon dioxide, so that there's very little loss of oxygen to make good by shipments from Earth. If they fail, there's a big enough reserve to last until the hotel can be evacuated or the plant repaired.

Almost as important is the heat-regulating apparatus. Out in space, in direct sunlight, an object can reach a temperature of three or four hundred degrees Fahrenheit on its "day" side, while the "night" side can be a couple of hundred degrees *below* zero. By circulating air through the double walls of the hotel, these temperature extremes are eliminated.

Ignoring such activities as poker and canasta, which are highly independent of gravity, there are two classes of recreation aboard the hotel. In the ring you can play most of the games that are found on Earth—with suitable modifications. The billiard tables, for example, have to be curved slightly: at first sight it looks as if they dip down in the middle, but in this radial gravity field, this makes them behave as flat surfaces. You very quickly get used to this sort of thing, though it may throw off your game for a while when you return to Earth.

However, since there's little point in going out into space to indulge in terrestrial-type sports, most of the excess energy in Sky Hotel is expended in the zero-gee rooms aboard the ball. The one thing that nobody misses is a chance to do some flying—*real* flying, of the kind we've all dreamed about at some time or another. You may feel a little foolish as you fasten the triangular wings between your ankles and wrists and secure the free ends to your belt. Certainly your first few strokes will start you turning helplessly over and over in the air. But in a few hours you'll be flying like a bird—and much less effortlessly. By the way, the crash helmet that goes with the wings is not just an ornament. It may prevent your knocking yourself out if you get up too much speed and don't notice how near the wall you are.

Some of the zero-gee ballets, with special lighting effects, that the expert performers can execute are unbelievably beautiful, like fairyland filmed in slow motion. Even if you've already seen them on TV, don't miss an opportunity to attend an actual performance at the hotel.

When you've earned your wings in the amusing series of tests that entitles you to your "Spacehound's Certificate," you'll probably want to take part in such sports as zero-gee basketball or three-dimensional miniature golf. Many terrestrial games have been adapted, with interesting variations, to conditions of weightlessness, but there are also dozens of sports and tricks that have no counterpart on Earth.

For example, there's the quite exhausting game you can play where everyone puts on wings and the winner is the one who can collect the largest number of scattered water drops into a single sphere—*and* brings it back to goal before his opponents tear it to pieces.

Talking about water drops leads me, inevitably, to the hotel's most incredible novelty—its famous swimming pool. Any resemblance to similarly described places on Earth is not merely coincidental—it's nonexistent.

When you go to the "pool" you'll find yourself in a big spherical chamber about sixty feet across, almost filled by what is claimed—probably correctly—to be the largest single drop of water in existence. You won't be particularly surprised to see people swimming around and around inside the sphere, but what *will* astonish you is the sight of a group in its center talking and laughing together and perhaps even taking refreshments. Even in space, you'll say to yourself, people still have to breathe!

To settle the mystery, dive into the drop and swim through it. When you've gone about twenty feet and are still some distance from the center, you'll break through another water surface and find yourself in a hollow space about ten feet across, breathing ordinary air. Yes, you're inside a bubble! It can't escape from the inside of the drop, because only when there is an "up" can bubbles rise in a liquid. So the swimming pool is really a huge hollow shell of water, and you can sit quietly at the very center and watch your friends sporting like fish all around you.

I *have* seen people smoking in the middle of the pool, though that's against regulations as it's liable to overtax the little air purifier that floats at the exact center of the bubble.

Incidentally, keeping the water clean presents some headaches, and you'll notice eight large pipes leading into the giant drop at points equally spaced over its surface. Water flows in and out of these at carefully adjusted rates, so that shell of liquid always remains the same size.

When you're tired of swimming, you can spend a good many happy hours in the observation lounge, simply watching the Earth and stars. There are no windows in the ring, because it would be rather disconcerting to see the heavens around you revolving at such a rate. So you'll have to do all your stargazing from the non-rotating ball.

From ten thousand miles out, the Earth is just small enough to fill your field of vision completely, and you can see everything except the extreme polar regions. Even to the naked eye, it's a source of endless enchantment. In the nine hours that the hotel takes to complete its orbit, you'll see the Earth change from new to full and back again, going through the phases that the Moon takes a whole month to complete. The sight of the dawn down there, as the Sun comes blasting up through the incandescent mists at the edge of the atmosphere and Earth grows swiftly from a hairline crescent to a huge glowing disk, is something no amount of repetition can ever stale.

When you've had your fill of gazing through the observation windows, you'll turn to the telescopes. Some of them can magnify up to a thousand times, so you'll feel that you're hanging only ten miles above the surface of the Earth. If there's no cloud, it's amazing how much minute detail you can see. Towns and cities are easy; even single large buildings can be detected under favorable conditions. But don't believe anyone who tells you that they've been able to see individual men! That's only possible from the inner satellites, a mere few hundred miles up.

It's interesting to study the effect of these novel surroundings on your companions. Human beings are incredibly adaptable, and for most of the time the guests in Sky Hotel enjoy themselves in the same uninhibited way as if they were down on Earth. But from time to time you'll catch them looking thoughtfully at the stars, realizing that *this* is space, *this* is the Universe. They'll have become suddenly aware that the familiar Earth, with its gravity and its air and its oceans, and its teeming, multitudinous life, is a freak, an incredible rarity; 99.999999 . . . per cent of the cosmos is emptiness and night.

That realization can affect people in two ways. It can depress them when they think how puny man is against the Universe, or it can exhilarate them when they consider his courage in attempting to conquer it.

Moving in almost exactly the same orbit as the hotel, but fifty miles away from it, is the newest and largest of the space hospitals —Haven IV. It's often possible to arrange a trip across in one of the low-powered rocket shuttles that ply between the orbits of the various stations, and sometimes there are official conducted tours of the hospital. Most of the patients on Haven are heart cases, recuperating under conditions where physical effort is much less than on Earth and their weakened hearts haven't got to pump pounds of blood up and down the body twenty-four hours a day. The first rocketeers, crushed in their acceleration couches under the strain of blast-off, would have been very surprised to know how soon cardiac sufferers were to make the same trip. Of course, the patients all travel under deep anesthesia and don't know a thing about it.

Haven IV is a single giant disk, slowly turning on its axis so that at the outer rim "gravity" has the same value as on Earth. As you go toward the center and the speed of rotation decreases, the synthetic gravity weakens as well, until at the very center you have complete weightlessness. New patients start their treatment near the axis of the hospital, in wards where gravity is maybe a tenth as powerful as it is on Earth, and move outward toward normal weight as their condition improves. Sometimes they never recover sufficiently to return to Earth, but even these severe cases can settle down on the Moon and get along happily with a sixth of Earth's gravity.

Besides the heart cases, the space hospitals specialize in polio victims, as well as people who have lost their legs and would be virtually helpless down on Earth. There are quite a number of legless men working permanently on the space stations. Often they are more agile than those who are not disabled—they haven't so much useless mass to drag around!

Quite recently, Haven IV has started to deal with severe burns. It doesn't take much imagination to realize how treatment and recuperation can be speeded when the patient can float freely in space and no longer has to lie on his dressings.

No wonder, therefore, that it's been said that the four space hospitals have already repaid humanity all the billions that the conquest of space has cost. And I haven't even mentioned the fundamental medical research they've made possible, particularly through the studies of giant microbes that could only be bred under zero-gee conditions.

From the observation lounge of the Sky Hotel you can see all the inner stations as they pass between you and Earth, moving on their smaller, swifter orbits far more rapidly than you do. Sometimes, when you are looking through a telescope at the lights of some city on the night side of Earth, you may be surprised to see

a tiny star explode against the darkness and start moving purpose-
fully out into space. You'll have caught one of the interplanetary
liners at the moment of take-off, as it pulls away from its refueling
station and begins its long journey. And sometimes you may see the
glare of one of the big freighter rockets as it starts the climb up
from Earth—that two-hundred-mile haul that requires so much
more effort than all the millions of miles between the planets.

Down there between you and Earth are the met stations, charting
the weather over the entire planet so that we know now, nine times
out of ten, exactly what's going to happen during the next forty-
eight hours. (The meteorologists are still worried about that odd
tenth time, but they swear they're going to get it licked one of these
days.) And there are the big space labs, carrying out all sorts of
experiments that could never be done on Earth, where no amount
of money could buy you a perfect vacuum as many miles across
as you cared to specify. Last, but perhaps most important of all,
are the astronomical observatories with their vast, floating mirrors,
scores of feet across, peering out across the billions of light-years
and no longer half-blinded by the murk and haze of the atmosphere.

You may feel rather superior to these lower satellites as you look
down upon them from your ten-thousand-mile-high eyrie. But if
you do, then remember that the outermost of all Earth's man-made
moonlets are twelve thousand miles beyond you. I mean, of course,
the three relay stations which now carry all the long-range TV and
radio traffic of the planet.

At this height of twenty-two thousand miles, a satellite takes
exactly twenty-four hours to go around its orbit, so the entire huge
triangle of the relay chain rotates in synchronism with Earth, just
as if it was fixed to it by invisible spokes. That's why, once you've
aimed your TV antenna at the nearest relay up there in the sky,
you need never move it again. And you can get your pictures without any interference, and from any spot in the world—something
that would have been incredible when TV was first invented.

Sometimes you can thumb a lift on a shuttle up to one of the
relay stations. Out there, more than twenty thousand miles above

the Earth, you'll really feel you're on the frontier of space. But don't forget that this is only a tenth of the distance to our nearest neighbor, the Moon, and much less than a thousandth of the distance to Mars or Venus, even at their closest approach. So when you get back to Earth, don't be too boastful about your achievements—at least until you've made quite sure that there are no *real* spacehounds in the party.

More seriously, there's one point you must watch when you're home again. Take things *very* easily for the first few days. Remember, we've got a little thing called gravity down here, and the tricks you can play in Sky Hotel won't work so well back on Earth. You can't cross Fifth Avenue, for instance, by stepping out at the two hundredth floor of Planet Tower and launching yourself in an easterly direction. (Take my word for it—it's been tried.) Even in your own home, you may find yourself treating the stairs with quite unjustified contempt, so this warning is by no means as superfluous as it seems.

Finally, I've been asked to deny a canard which has been causing the hotel management much grief. Luigi, *chef de cuisine,* is particularly upset by the slander that he's convinced has been put out by rival establishments on Earth. It's completely untrue that the guests at the hotel have to live on compressed foods and vitamin pills, like the first space pioneers. The meals are as good as anything you can get on Earth. They may not actually *weigh* as much, but I can assure you, from personal experience, that they're every bit as satisfying.

So You're Going to Mars?

$$\underline{\mathbf{6}}$$

This essay was written in 1952, long before the Mariner space probes gave us our close-up glimpses of the tantalizing red planet. Nevertheless, most of the concepts presented here are still quite valid, though we now know that Mars is even more rugged than anticipated. In particular, the atmospheric pressure is so low (about one-hundredth of Earth's) that simple breathing masks will not give sufficient protection; we will have to wear spacesuits.

Many of the ideas in this article were worked out in much more detail in my novel *The Sands of Mars.*

So you're going to Mars? That's still quite an adventure— though I suppose that in another ten years no one will think twice about it. Sometimes it's hard to remember that the first ships reached Mars scarcely more than half a century ago and that our colony on the planet is less than thirty years old. (By the way, don't use *that* word when you get there. Base, settlement, or whatever you like—but not colony, unless you want to hear the ice tinkling all around you.)

I suppose you've read all the forms and tourist literature they gave you at the Department of Extraterrestrial Affairs. But there's a lot you won't learn just by reading, so here are some pointers and background information that may make your trip more enjoyable.

I won't say it's right up to date—things change so rapidly, and it's a year since I got back from Mars myself—but on the whole you'll find it pretty reliable.

Presumably you're going just for curiosity and excitement—because you want to see what life is like out on the new frontier. It's only fair, therefore, to point out that most of your fellow passengers will be engineers, scientists, or administrators traveling to Mars—some of them not for the first time—because they've got a job of work to do. So whatever your achievements here on Earth, it's advisable not to talk too much about them, as you'll be among people who've had to tackle much tougher propositions. I won't say that you'll find them boastful: it's simply that they've got a lot to be proud of, and they don't mind who knows it.

If you haven't booked your passage yet, remember that the cost of the ticket varies considerably according to the relative positions of Mars and Earth. That's a complication we don't have to worry about when we're traveling from country to country on our own globe, but Mars can be six times farther away at one time than at another. Oddly enough, the shortest trips are the most expensive, since they involve the greatest changes of speed as you hop from one orbit to the other. And in space, speed, not distance, is what costs money.

Incidentally, I'd like to know how you've managed it. I believe the cheapest round trip comes to about $30,000, and unless the firm is backing you or you've got a very elastic expense account—Oh, all right, if you don't want to talk about it . . .

I take it you're O.K. on the medical side. That examination isn't for fun, nor is it intended to scare anyone off. The physical strain involved in space flight is negligible—but you'll be spending at least two months on the trip, and it would be a pity if your teeth or your appendix started to misbehave. See what I mean?

You're probably wondering how you can possibly manage on the weight allowance you've got. Well, it can be done. The first thing to remember is that you don't need to take any suits. There's no weather inside a spaceship; the temperature never varies more than a couple of degrees over the whole trip, and it's held at a fairly high value so that all you'll want is an ultra-lightweight tropical kit. When you get to Mars you'll buy what you need there and dump it when you return. The great thing to remember is *only carry the stuff you actually need on the trip.* I strongly advise you to buy one of the complete travel kits—a store like Abercrombie & Fitch can supply the approved outfits. They're expensive, but will save you money on excess baggage charges.

Take a camera by all means—there's a chance of some unforgettable shots as you leave Earth and when you approach Mars. But there's nothing to photograph on the voyage itself, and I'd advise you to take all your pictures on the outward trip. You can sell a good camera on Mars for five times its price here—and save yourself the cost of freighting it home. They don't mention *that* in the official handouts.

Now that we've brought up the subject of money, I'd better remind you that the Martian economy is quite different from anything you'll meet on Earth. Down here, it doesn't cost you anything to breathe, even though you've got to pay to eat. But on Mars the very air has to be synthesized—they break down the oxides in the ground to do this—so every time you fill your lungs someone has to foot the bill. Food production is planned in the same way—each of the cities, remember, is a carefully balanced ecological system, like a well-organized aquarium. No parasites can be allowed, so everyone has to pay a basic tax which entitles him to air, food, and the shelter of the domes. The tax varies from city to city, but averages about $10 a day. Since everyone earns at least ten times as much as this, they can all afford to go on breathing.

You'll have to pay this tax, of course, and you'll find it rather hard to spend much more money than this. Once the basic needs for life are taken care of, there aren't many luxuries on Mars. When they've got used to the idea of having tourists around, no doubt they'll get organized, but as things are now you'll find that most reasonable requests won't cost you anything. However, I should make arrangements to transfer a substantial credit balance to the

Bank of Mars—if you've still got anything left. You can do that by radio, of course, before you leave Earth.

So much for the preliminaries; now some points about the trip itself. The ferry rocket will probably leave from the New Guinea field, which is about two miles above sea level on the top of the Orange Range. People sometimes wonder why they chose such an out-of-the-way spot. That's simple: it's on the equator, so a ship gets the full thousand-mile-an-hour boost of the Earth's spin as it takes off—and there's the whole width of the Pacific for jettisoned fuel tanks to fall into. And if you've ever *heard* a spaceship taking off, you'll understand why the launching sites have to be a few hundred miles from civilization.

Don't be alarmed by anything you've been told about the strain of blast-off. There's really nothing to it if you're in good health— and you won't be allowed inside a spaceship unless you are. You just lie down on the acceleration couch, put in your earplugs, and relax. It takes over a minute for the full thrust to build up, and by that time you're quite accustomed to it. You'll have some difficulty in breathing, perhaps—it's never bothered me—but if you don't attempt to move you'll hardly feel the increase of weight. What you *will* notice is the noise, which is slightly unbelievable. Still, it lasts only five minutes, and by the end of that time you'll be up in the orbit and the motors will cut out. Don't worry about your hearing; it will get back to normal in a couple of hours.

You won't see a great deal until you get aboard the space station, because there are no viewing ports on the ferry rockets and passengers aren't encouraged to wander around. It usually takes about thirty minutes to make the necessary steering corrections and to match speed with the station; you'll know when that's happened from the rather alarming "clang" as the air locks make contact. Then you can undo your safety belt, and of course you'll want to see what it's like being weightless.

Now, take your time, and do exactly what you're told. Hang on to the guide rope through the air lock and don't try to go flying around like a bird. There'll be plenty of time for that later: there's not enough room in the ferry, and if you attempt any of the usual tricks you'll not only injure yourself but may damage the equipment as well.

Space Station One, which is where the ferries and the liners meet to transfer their cargoes, takes just two hours to make one circuit of the Earth. You'll spend all your time in the observation lounge: everyone does, no matter how many times they've been out into space. I won't attempt to describe that incredible view; I'll merely remind you that in the hundred and twenty minutes it takes the station to complete its orbit you'll see the Earth wax from a thin crescent to a gigantic, multicolored disk, and then shrink again to a black shield eclipsing the stars. As you pass over the night side you'll see the lights of cities down there in the darkness, like patches of phosphorescence. And the stars! You'll realize that you've never really seen them before in your life.

But enough of these purple passages; let's stick to business. You'll probably remain on Space Station One for about twelve hours, which will give you plenty of opportunity to see how you like weightlessness. It doesn't take long to learn how to move around; the main secret is to avoid all violent motions—otherwise you may crack your head on the ceiling. Except, of course, that there isn't a ceiling since there's no up or down any more. At first you'll find that confusing: you'll have to stop and decide which direction you want to move in, and then adjust your personal reference system to fit. After a few days in space it will be second nature to you.

Don't forget that the station is your last link with Earth. If you want to make any final purchases, or leave something to be sent home—do it then. You won't have another chance for a good many million miles. But beware of buying items that the station shop assures you are "just the thing on Mars."

You'll go aboard the liner when you've had your final medical check, and the steward will show you to the little cabin that will be your home for the next few months. Don't be upset because you can touch all the walls without moving from one spot. You'll only have to sleep there, after all, and you've got the rest of the ship to stretch your legs in.

If you're on one of the larger liners, there'll be about a hundred other passengers and a crew of perhaps twenty. You'll get to know them all by the end of the voyage. There's nothing on Earth quite like the atmosphere in a spaceship. You're a little, self-contained community floating in vacuum millions of miles from anywhere, kept alive in a bubble of plastic and metal. If you're a good mixer, you'll find the experience very stimulating. But it has its disadvantages. The one great danger of space flight is that some prize bore may get on the passenger list—and short of pushing him out of the air lock there's nothing anyone can do about it.

It won't take you long to find your way around the ship and to get used to its gadgets. Handling liquids is the main skill you'll have to acquire: your first attempts at drinking are apt to be messy. Oddly enough, taking a shower is quite simple. You do it in sort of plastic cocoon, and a circulating air current carries the water out at the bottom.

At first the absence of gravity may make sleeping difficult— you'll miss your accustomed weight. That's why the sheets over the bunks have spring tensioning. They'll keep you from drifting out while you sleep, and their pressure will give you a spurious sensation of weight.

But learning to live under zero gravity is something one can't be taught in advance: you have to find out by experience and practical demonstration. I believe you'll enjoy it, and when the novelty's worn off you'll take it completely for granted. Then the problem will be getting used to gravity again when you reach Mars!

Unlike the take-off of the ferry rocket from Earth, the breakaway of the liner from its satellite orbit is so gentle and protracted that it lacks all drama. When the loading and instrument checks have been completed, the ship will uncouple from the Space Station and drift a few miles away. You'll hardly notice it when the atomic drive goes on; there will be the faintest of vibrations and a feeble sensation of weight. The ship's acceleration is so small, in fact, that you'll weigh only a few ounces, which will scarcely interfere with your freedom of movement at all. Its only effect will be to make things drift slowly to one end of the cabin if they're left lying around.

Although the liner's acceleration is so small that it will take hours to break away from Earth and head out into space, after a week of continuous drive the ship will have built up a colossal speed. Then the motors will be cut out and you'll carry on under your own momentum until you reach the orbit of Mars and have to start thinking about slowing down.

Whether your weeks in space are boring or not depends very much on you and your fellow passengers. Quite a number of entertainments get organized on the voyage, and a good deal of money is liable to change hands before the end of the trip. (It's a curious fact, but the crew usually seems to come out on top.) You'll have plenty of time for reading, and the ship will have a good library of microbooks. There will be radio and TV contact with Earth and Mars for the whole voyage, so you'll be able to keep in touch with things—if you want to.

On my first trip, I spent a lot of my time learning my way around the stars and looking at clusters and nebulae through a small telescope I borrowed from the navigation officer. Even if you've never felt the slightest interest in astronomy before, you'll probably be a keen observer before the end of the voyage. Having the stars all around you—not merely overhead—is an experience you'll never forget.

As far as outside events are concerned, you realize, of course, that absolutely nothing can happen during the voyage. Once the drive has cut out, you'll seem to be hanging motionless in space: you'll be no more conscious of your speed than you are of Earth's seventy thousand miles an hour around the Sun right now. The only evidence of your velocity will be the slow movement of the nearer planets against the background of the stars—and you'll have to watch carefully for a good many hours before you can detect even this.

By the way, I hope you aren't one of those foolish people who are still frightened about meteors. They see that enormous chunk of nickel-steel in New York's American Museum of Natural History and imagine that's the sort of thing you'll run smack into as soon as you leave the atmosphere—forgetting that there's rather a lot of room in space and that even the biggest ship is a mighty small target. You'd have to sit out there and wait a good many centuries before a meteor big enough to puncture the hull came along. It hasn't happened to a spaceship yet.

One of the big moments of the trip will come when you realize that Mars has begun to show a visible disk. The first feature you'll be able to see with the naked eye will be one of the polar caps, glittering like a tiny star on the edge of the planet. A few days later the dark areas—the so-called seas—will begin to appear, and presently you'll glimpse the prominent triangle of the Syrtis Major. In the week before landing, as the planet swims nearer and nearer, you'll get to know its geography pretty thoroughly.

The braking period doesn't last very long, as the ship has lost a good deal of its speed in the climb outward from the Sun. When it's over you'll be dropping down onto Phobos, the inner moon of Mars, which acts as a natural space station about four thousand miles above the surface of the planet. Though Phobos is only a jagged lump of rock not much bigger than some terrestrial mountains, it's reassuring to be in contact with something solid again after so many weeks in space.

When the ship has settled down into the landing cradle, the air lock will be coupled up and you'll go through a connecting tube into the port. Since Phobos is much too small to have an appreciable gravity, you'll still be effectively weightless. While the ship's being unloaded the immigration officials will check your papers. I don't know the point of this; I've never heard of anyone being sent all the way back to Earth after having got this far!

There are two things you mustn't miss at Port Phobos. The restaurant there is quite good, even though the food is largely synthetic; it's very small, and only goes into action when a liner docks, but it does its best to give you a fine welcome to Mars. And after a couple of months you'll have got rather tired of the shipboard menu.

The other item is the centrifuge; I believe that's compulsory now. You go inside and it will spin you up to half a gravity, or rather more than the weight Mars will give you when you land. It's simply a little cabin on a rotating arm, and there's room to walk around inside so that you can practice using your legs again. You probably won't like the feeling; life in a spaceship can make you lazy.

The ferry rockets that will take you down to Mars will be waiting when the ship docks. If you're unlucky you'll hang around at the port for some hours, because they can't carry more than twenty passengers and there are only two ferries in service. The actual descent to the planet takes about three hours, and it's the only time on the whole trip when you'll get any impression of speed. Those ferries enter the atmosphere at over five thousand miles an hour and go halfway around Mars before they lose enough speed through air resistance to land like ordinary aircraft.

You'll land, of course, at Port Lowell: besides being the largest settlement on Mars it's still the only place that has the facilities for handling spaceships. From the air the plastic pressure domes look like a cluster of bubbles—a very pretty sight when the Sun catches them. Don't be alarmed if one of them is deflated. That doesn't mean that there's been an accident. The domes are let down at fairly frequent intervals so that the envelopes can be checked for leaks. If you're lucky you may see one being pumped up—it's quite impressive.

After two months in a spaceship, even Port Lowell will seem a mighty metropolis. (Actually, I believe its population is now well over twenty thousand.) You'll find the people energetic, inquisitive, forthright—and very friendly, unless they think you're trying to be superior.

It's a good working rule never to criticize anything you see on Mars. As I said before, they're very proud of their achievements—and after all you *are* a guest, even if a paying one.

Port Lowell has practically everything you'll find in a city on Earth, though of course on a smaller scale. You'll come across many reminders of "home." For example, the main street in the city is Fifth Avenue—but surprisingly enough you'll find Piccadilly Circus where it crosses Broadway.

The port, like all the major settlements, lies in the dark belt of vegetation that roughly follows the Equator and occupies about half the southern hemisphere. The northern hemisphere is almost all desert—the red oxides that give the planet its ruddy color. Some of these desert regions are very beautiful; they're far older than anything on the surface of our Earth, because there's been little weathering on Mars to wear down the rocks—at least since the seas dried up, more than 500 million years ago.

You shouldn't attempt to leave the city until you've become quite accustomed to living in an oxygen-rich, low-pressure atmosphere. You'll have grown fairly well acclimatized on the trip, because the air in the spaceship will have been slowly adjusted to conditions on Mars. Outside the domes, the pressure of the natural Martian atmosphere is about equal to that on the top of Mount Everest—and it contains practically no oxygen. So when you go out you'll have to wear a helmet, or travel in one of those pressurized jeeps they call "sand fleas."

Wearing a helmet, by the way, is nothing like the nuisance you'd expect it to be. The equipment is very light and compact and, as long as you don't do anything silly, is quite foolproof. As it's very unlikely that you'll ever go out without an experienced guide, you'll have no need to worry. Thanks to the low gravity, enough oxygen for twelve hours' normal working can be carried quite easily—and you'll never be away from shelter as long as that.

Don't attempt to imitate any of the locals you may see walking around without oxygen gear. They're second-generation colonists and are used to the low pressure. They can't breathe the Martian atmosphere any more than you can, but like the old-time native pearl divers they can make one lungful last for several minutes when necesary. Even so, it's a silly sort of trick and they're not supposed to do it.

As you know, the other great obstacle to life on Mars is the low temperature. The highest thermometer reading ever recorded is somewhere in the eighties, but that's quite exceptional. In the long winters, and during the night in summer *or* winter, it never rises above freezing. And I believe the record low is minus one hundred and ninety!

Well, you won't be outdoors at night, and for the sort of excursions you'll be doing, all that's needed is a simple thermosuit. It's very light, and traps the body heat so effectively that no other source of warmth is needed.

No doubt you'll want to see as much of Mars as you can during your stay. There are only two methods of transport outside the cities—sand fleas for short ranges and aircraft for longer distances. Don't misunderstand me when I say "short ranges"—a sand flea with a full charge of power cells is good for a couple of thousand miles, and it can do eighty miles an hour over good ground. Mars could never have been explored without them. You can *survey* a planet from space, but in the end someone with a pick and shovel has to do the dirty work filling in the map.

One thing that few visitors realize is just how big Mars is. Although it seems small beside the Earth, its land area is almost as great because so much of our planet is covered with oceans. So it's hardly surprising that there are vast regions that have never been properly explored, particularly around the poles. Those stubborn people who still believe that there was once an indigenous Martian civilization pin their hopes on these great blanks. Every so often you hear rumors of some wonderful archaeological discovery in the wastelands, but nothing ever comes of it.

Personally, I don't believe there ever *were* any Martians—but the planet is interesting enough for its own sake. You'll be fascinated by the plant life and the queer animals that manage to live without oxygen, migrating each year from hemisphere to hemisphere, across the ancient sea beds, to avoid the ferocious winter.

The fight for survival on Mars has been fierce, and evolution has produced some pretty odd results. Don't go investigating any Martian life forms unless you have a guide, or you may get some unpleasant surprises. Some plants are so hungry for heat that they may try to wrap themselves around you.

Well, that's all I've got to say, except to wish you a pleasant trip. Oh, there *is* one other thing. My boy collects stamps, and I rather let him down when I was on Mars. If you could drop me a few letters while you're there—there's no need to put anything in them if you're too busy—I'd be much obliged. He's trying to collect a set of space-mail covers postmarked from each of the principal Martian cities, and if you could help—thanks a lot!

Next—The Planets!

7

This paper was presented as an address to the Fourth International Symposium on Bioastronautics and the Exploration of Space, arranged by the Aerospace Medical Division, Brooks Air Force Base, San Antonio, Texas, in June 1968. The other participants included Dr. Edward Welsh, Dr. Fred Whipple, Dr. Harold Urey, Dr. Cyril Ponnamperuma, Dr. Fritz Zwicky, Dr. Robert Gilruth, Dr. Charles Berry, Dr. Krafft Ehricke, Dr. Willard Libby—and the Dean of Space Medicine, Dr. Hubertus Strughold. I have fond recollections of this meeting, as apart from the good company and the Air Force hospitality, it also provided several bonuses—the excellent "Hemisfair," the Alamo, and my first stunned encounter with that piece of pure James Bondage, the Bell one-man Rocket-Belt.

The possibility of life on Venus now appears even more remote than when this paper was presented; the surface of the planet may justifiably be described as an inferno. But Venus has surprised us many times before. . . .

The suggestion, at the end of the essay, that we are observing new sources of energy which may far exceed those of the atomic nucleus now seems more and more probable, thanks to recent researches on "quasars." The Universe can provide all the power we shall ever need to drive *real* star ships—if we are clever enough to tap it. When that time comes, let us hope that our cleverness does not, as it does today, exceed our wisdom.

◎

It has been said that history never repeats itself, but that historical situations recur. To anyone who, like myself, has been involved in astronautical activities for over thirty years, there must be a feeling of familiarity, of "I have been here before," in some of the present arguments about the exploration of space.

Like all revolutionary new ideas, the subject has had to pass through three stages, which may be summed up by these reactions: (1) "It's crazy—don't waste my time." (2) "It's possible, but it's not worth doing." (3) "I always said it was a good idea."

As far as orbital flights, and even journeys to the Moon, are concerned, we have made good progress through all these stages, though it will be a few years yet before everyone is in category 3. But where flights to the planets are involved, we are still almost where we were thirty years ago. True, there is much less complete skepticism—to that extent, history has *not* repeated itself—but there remains, despite all the events of the past decade, a widespread misunderstanding of the possible scale, importance, and ultimate implications of travel to the planets.

Let us start by looking at some fundamentals, which are not as well known as they should be, even to space scientists. Forgetting all about rockets and today's astronautical techniques, consider the basic problem of lifting a man away from the Earth, purely in terms of the work done to move him against gravity.

For a man of average mass, the energy requirement is about 1000 kilowatt-hours, which customers with a favorable tariff can purchase for $10 from their electric utility company. *What may be called the basic cost of a one-way ticket to space is thus the modest sum of $10.*

For the smaller planets and all satellites—Mercury, Venus, Mars, Pluto, Moon, Titan, Ganymede, etc.—the exit fee is even less; you need only 50 cents worth of energy to escape from the Moon. Giant planets like Jupiter, Saturn, Uranus, and Neptune are naturally much more expensive propositions. If you are ever stranded on Jupiter, you'll have to buy almost $300 of energy to get home. Make sure you have enough traveler's checks. . . .

Of course, the planetary fields are only part of the story; work also has to be done traveling from orbit to orbit, and thus moving up or down the enormous gravitational field of the Sun. But by great good luck the Solar System appears to have been designed for the convenience of space travelers: all the planets lie far out on the gentle slope of the solar field where it merges into the endless plain of interstellar space. In this respect, the conventional map of the Solar System, showing the planets clustering around the Sun, is wholly misleading.

We can say, in fact, that the planets are 99 per cent free of the Sun's gravitational field, so that the energy required for orbital transfers is quite small; usually it is considerably less than that needed to escape from the planets themselves. In dollars and cents, the energy cost of transferring a man from the surface of the Earth to that of Mars is less than $20. Even for the worst possible case (surface of Jupiter to surface of Saturn), the pure energy cost is less than $1,000!

Hardheaded rocket engineers may well consider that the above arguments, purporting to prove that space travel should be about a billion times cheaper than it is, have no relevance to the practical case—since even today the cost of the fuel is trivial compared with the cost of the hardware. Most of the mountainous Saturn 5 standing on the pad can be bought for, quite literally, a few cents a pound; that is all that kerosene and liquid oxygen cost. The expensive items are the precision-shaped pieces of high-grade metals, and all the little black boxes that are sold by the carat.

Although this is true, it is also to a large extent a consequence of our present immature, no-margin-for-error technology. Just ask yourself how expensive driving would be, if a momentary engine failure was liable to write off your car—and yourself—and the fuel supply was so nicely calculated that you couldn't complete a mission if the parking meter you'd aimed at happened to be already occupied. This is roughly the situation for planetary travel today.

To imagine what it may one day become, let us look at the record of the past, and see what lessons we can draw from the early history of aeronautics. Soon after the failure of Langley's "Aerodrome" in 1903, the great astronomer Simon Newcomb wrote a famous essay, well worth rereading, which proved that heavier-than-air flight was impossible by means of known technology. The ink was hardly dry on the paper when a pair of bicycle mechanics irreverently threw grave doubt on the professor's conclusions. When informed of the embarrassing fact that the Wright brothers had just flown, Newcomb gamely replied, "Well, maybe a flying machine *can* be built. But it certainly couldn't carry a passenger as well as a pilot."

Now I am not trying to poke fun at one of the greatest of American scientists. When you look at the Wright biplane, hanging up there in the Smithsonian Institution, Newcomb's attitude seems very reasonable indeed; I wonder how many of us would have been prepared to dispute it in 1903.

Yet—and this is the really extraordinary point—there is a smooth line of development, *without any major technological breakthrough,* from the Wright "Flyer" to the last of the great piston-engined aircraft such as the DC-6. All the many-orders-of-magnitude improvement in performance came as a result of engineering advances which in retrospect seem completely straightforward, and sometimes even trivial. Let us list the more important ones: variable-pitch airscrews; slots and flaps; retractable undercarriages; concrete runways; streamlining; supercharging.

Not very spectacular, are they? Yet these things, together with steady improvements in materials and design, lifted much of the commerce of mankind into the air. For they had a synergistic effect on performance; their cumulative effect was much greater than could have been predicted by considering them individually. They did not merely add; they multiplied.

All this took about forty years; and then there was the second technological breakthrough—the advent of the jet engine—and a new cycle of development started.

Unless the record of the past is wholly misleading, we are going to see much the same sequence of events in space. As far as can be judged at the moment, the equivalent items on the table of aerospace progress may be: refueling in orbit; air-breathing boosters; reusable boosters; refueling on (or from) the Moon; lightweight materials (e.g., composites and fibers).

Probably the exploitation of these relatively conventional ideas will take somewhat less than the forty years needed in the case of aircraft; their full impact should be felt by the turn of the century. Well before then, moreover, the next breakthrough or quantum jump in space technology should also have occurred, with the development of new propulsion systems—presumably fission-powered but perhaps even using fusion as well.

And with these, the Solar System will become an extension of the Earth—if we wish it to be so.

It is at this point, however, that all analogy with the past breaks down; we can no longer draw meaningful parallels between aeronautics and astronautics. As soon as aircraft were shown to be practical, there were obvious and immensely important uses for them: military, commercial, scientific. They could be used to provide swifter connections between already highly developed communities —a state of affairs which almost certainly does not exist in the Solar System, and may not do so for centuries to come.

It seems, therefore, that we may be involved in a peculiarly vicious circle. Planetary exploration will not be really practical until we have developed a mature spaceship technology, but we won't have good spaceships until we have worthwhile places to send them. Places, above all, with those adequate refueling and servicing facilities now sadly lacking elsewhere in the Solar System.

How can we escape from this dilemma? Fortunately, there is one encouraging factor.

Almost the whole of the technology needed for long-range space travel will, inevitably and automatically, be developed during the exploitation of *near* space. Even if we set our sights no higher than a thousand miles above the Earth, we would find that by the time we had perfected the high-thrust, high-performance surface-to-surface transports, the low-acceleration interorbital shuttles, the reliable, closed-cycle space-station ecologies, we would have proved out at least 90 per cent of the technology needed for the exploration of the Solar System. And the most expensive 90 per cent at that. . . .

Perhaps I had better spend a few moments here on those strange characters who think that space is the exclusive province of automatic robot probes and that we should stay at home and watch TV, as God intended us to. This whole man-machine controversy will seem, in another couple of decades, to be a baffling mental aberration of the Early Space Age.

I won't waste any time arguing with this viewpoint, as I hold these truths to be self-evident: (1) unmanned spacecraft should be used whenever they can do a job more efficiently, cheaply, and safely than manned vehicles; (2) until we have automatons superior to human beings (by which time all bets will be off), all really sophisticated space operations will demand human participation. I refer to such activities as assembling and servicing the giant applications satellites of the next decade; running orbital observatories, laboratories, hospitals, factories—projects for which there will be such obvious and overwhelming commercial and scientific benefits that no one will dispute them.

In particular, the impact on Solar System studies of medium-sized telescopes outside the atmosphere—a mere couple of hundred miles above the Earth!—will be overwhelming. Until the advent of radar and space probes, everything we knew about the planets had been painfully gathered, over a period of about a century and a half, by astronomers with inadequate instruments, hastily sketching details on a tiny, trembling disk glimpsed during moments of good visibility. Such moments—when the atmosphere is stable and the image undistorted—may add up to only a few hours in an entire lifetime of observing.

In these circumstances, it would be amazing if we had acquired any *reliable* knowledge about planetary conditions; it is safest to assume that we have not. We are still in the same position as the medieval cartographers with their large areas of "Terra Incognita" and their "Here Be Dragons," except that we may have gone too far in the other direction—"Here Be *No* Dragons." Our ignorance is so great that we have no right to make either assumption.

As proof of this, let me remind you of some horrid shocks the astronomers have received recently, when things of which they were quite sure turned out to be simply not true. The most embarrassing example is the rotation of Mercury: until a couple of years ago, everyone was perfectly certain that it always kept the same face toward the Sun, so that one side was eternally dark, the other eternally baked. But now, radar observations indicate that it turns on its axis every fifty-nine days; it has sunrise and sunset like any respectable world. Nature seems to have played a dirty trick on several generations of patient astronomers.

Einstein once said, "The good Lord is subtle, but He is not malicious." The case of Mercury casts some doubt on this dictum. And what about Venus? You can find, in the various reference books, rotation periods for Venus ranging all the way from twenty-four hours up to the full value of the year, 225 days. But, as far as I know, not one astronomer ever suggested that Venus would present the extraordinary case of a planet with a day longer than its year! And, of course, it *would* be the one example we had no way of checking, until the advent of radar. Is this subtlety—or malice?

And look at the Moon. Five years ago, everyone was certain that its surface was either soft dust or hard lava. If the two schools of thought had been on speaking terms, they would at least have agreed that there were no alternatives. But then Luna 9 and Surveyor 1 landed and what did they find? Good honest dirt. . . .

These are by no means the only examples of recent shocks and surprises. There's the unexpectedly high temperature beneath the clouds of Venus; the craters of Mars; the gigantic radio emissions from Jupiter; the complex organic chemicals in certain meteors;

the clear signs of extensive activity on the surface of the Moon. And now Mars seems to be turning inside out. The ancient, dried-up sea beds may be as much a myth as Dejah Thoris, Princess of Helium; for it looks as if the dark *maria* are actually highlands, not lowlands, as we had always thought.

The negative point I am making is that we really know nothing about the planets. The positive one is that a tremendous amount of reconnaissance—the essential prelude to *manned* exploration—can be carried out from Earth orbit. It is probably no exaggeration to say that a good orbiting telescope could give us a view of Mars at least as clear as did Mariner 4. And it would be a view infinitely more valuable—a continuous coverage of the whole visible face, not a single snapshot of a small percentage.

Nevertheless, there are many tasks which can best be carried out by unmanned spacecraft. Among these is one which, though of great scientific value, is of even more profound psychological importance. I refer to the production of low-altitude oblique photographs.

It is no disparagement of the wonderful Ranger, Luna, and Surveyor coverage to remind you that what suddenly made the Moon a real place, and not merely an astronomical body up there in the sky, was the famous photograph of Copernicus from Lunar Orbiter 2. When the newspapers called it the picture of the century, they were expressing a universally felt truth. This was the photograph that first proved to our emotions what our minds already knew but had never really believed—that Earth is not the only world. The first high-definition, oblique photos of Mars, Mercury, and the satellites of the giant planets will have a similar impact, bringing our mental images of these places into sharp focus for the first time.

The old astronomical writers had a phrase that has gone out of fashion but which may well be revived: the plurality of worlds. Yet, of course, every world is itself a plurality. To realize this, one has only to ask oneself: How long will it be before we have learned everything that can be known about the planet Earth? It will be quite a few centuries yet before terrestrial geology, oceanography, and geophysics are closed, "surprise-free" subjects.

Consider the multitude of environments that exists here on Earth, from the summit of Everest to the depths of the Marianas Trench—from high noon in Death Valley to midnight at the South Pole. We may have equal variety on the other planets, with all that this implies for the existence of life. It is amazing how often this elementary fact is overlooked and how often a single observation or even a single extrapolation from a preliminary observation based on a provisional theory has been promptly applied to a whole world.

It is possible, of course, that the Earth has a greater variety of more complex environments than any other planet. Like a jet-age tourist "doing Europe" in a week, we may be able to wrap up Mars or Venus with a relatively small number of "landers." But I doubt it, if only for the reason that the whole history of astronomy teaches us to be cautious of any theory purporting to show that there is something special about the Earth. In their various ways, the other planets may have orders of complexity as great as ours. Even the Moon—which seemed a promising candidate for geophysical simplicity less than a decade ago—has already begun an avalanche of surprises.

The late Professor J. B. S. Haldane once remarked—and this should be called Haldane's Law—"The Universe is not only queerer than we imagine; it is queerer than we *can* imagine." We will encounter the operation of this law more and more frequently, as we move away from home. And as we prepare for this move, it is high time that we face up to one of the more shattering realities of the astronomical situation. For all practical purposes, we are still as geocentrically minded as if Copernicus had never been born; to all of us, the Earth is the center, if not of the Universe, at least of the Solar System.

Well, I have news for you. There is really only one planet that matters, and that planet is not Earth, but Jupiter. My esteemed colleague Isaac Asimov summed it up very well when he remarked, "The Solar System consists of Jupiter plus debris." Even spectacular Saturn doesn't count; he has less than a third of Jupiter's enormous mass—and Earth is a hundred times smaller than Saturn! Our planet is an unconsidered trifle, left over after the main building operations were completed.

This is quite a blow to our pride, but there may be much worse to come, and it is well to get ready for it. Jupiter may also be the *biological,* as well as the physical, center of gravity of the Solar System.

This, of course, represents a complete reversal of views in a couple of decades. Not long ago, it was customary to laugh at the naïve ideas of the early astronomers—Sir John Herschel, for example—who took it for granted that all the planets were teeming with life. This attitude is certainly overoptimistic, but it no longer seems as simple-minded as the opinion, to be found in the popular writings of the 1930s, that ours might be the only solar system, and, hence, the only abode of life in the entire Galaxy.

The pendulum has indeed swung—perhaps for the last time, for in another few decades we should know the truth. The discovery that Jupiter is quite warm, and has precisely the type of atmosphere in which life is believed to have arisen on Earth, may be the prelude to the most significant biological findings of this century. Carl Sagan and Jack Leonard put it well in their book, *Planets:* "Recent work on the origin of life and the environment of Jupiter suggests that it may be more favorable to life than any other planet, *not excepting the Earth.*" (My italics.)

The extraordinary color changes in the Jovian atmosphere—in particular, the behavior of that Earth-sized, drifting apparition the Great Red Spot—hint at the production of organic materials in enormous quantities. Where this happens, life may follow inevitably, given a sufficient lapse of time. To quote Isaac Asimov again, "If there are seas on Jupiter . . . think of the fishing."

So *that* may explain the mysterious disappearances and reappearances of the Great Red Spot. It is, as Polonius agreed in a slightly different context, "very like a whale."

Dr. James Edson, late of NASA, once remarked, "Jupiter is a

problem for my grandchildren." I suspect that he may have been wildly optimistic. The zoology of a world outweighing three hundred Earths could be the full-time occupation of mankind for the next thousand years.

It also appears that Venus, with its extremely dense, furnace-hot atmosphere, may be an almost equally severe yet equally promising challenge. There now seems little doubt that the planet's *average* temperature is around 700 degrees Fahrenheit; but this does not, as many have prematurely assumed, rule out all possibility of life—even life of the kind that exists on Earth.

There may be little mixing of the atmosphere, and hence little exchange of heat between the poles and the equator on a planet that revolves as slowly as Venus. At high latitudes, or great altitudes—and Venusian mountains have now been detected by radar—it may be cool enough for liquid water to exist. (Even on Earth, remember, the temperature difference between the hottest and the coldest points is almost 300 degrees.) What makes this more than idle speculation is the exciting discovery, by the Russian space probe Venera 5, of oxygen in the planet's atmosphere. This extremely reactive gas combines with so many materials that it cannot occur in the free state—unless it is continually renewed by vegetation. Free oxygen is an almost infallible indicator of life: if I may be allowed the modest cough of the minor prophet, I developed precisely this argument some years ago in a story of Venusian exploration, "Before Eden."

On the other hand, it is also possible that we shall discover no trace of extraterrestrial life, past or present, on any of the planets. This will be a great disappointment, but even such a negative finding would give us a much sounder understanding of the conditions in which living creatures are likely to evolve, and this in turn would clarify our views on the distribution of life in the universe as a whole. However, it seems much more probable that long before we can certify the Solar System as sterile, the communications engineers will have settled this ancient question—in the affirmative.

For that is what the exploration of space is really all about, and this is why many people are afraid of it, though they may give other reasons, even to themselves. It may be just as well that there are no contemporary higher civilizations in our immediate vicinity; the cultural shock of direct contact might be too great for us to survive. But by the time we have cut our teeth on the Solar System, we should be ready for such encounters. The challenge, in the Toynbeean sense of the word, should then bring forth the appropriate response.

Do not for a moment doubt that we will one day head out for the stars—if, of course, the stars do not reach us first. I think I have read most of the arguments proving that interstellar travel is impossible; they are latter-day echoes of Professor Newcomb's paper on heavier-than-air flight. The logic and the mathematics are impeccable; the premises wholly invalid. The more sophisticated are roughly equivalent to proving that dirigibles cannot break the sound barrier.

In the opening years of this century, the pioneers of astronautics were demonstrating that flight to the Moon and nearer planets was possible, though with great difficulty and expense, by means of chemical propellents. But even then, they were aware of the promise of nuclear energy and hoped that it would be the ultimate solution. They were right.

Today, it can likewise be shown that various conceivable, though currently quite impracticable, applications of nuclear and medical techniques could bring at least the closer stars within the range of exploration. And I would warn any skeptics who may point out the marginal nature of these techniques that at this very moment there are appearing simultaneously, on the twin horizons of the infinitely large and the infinitely small, unmistakable signs of a breakthrough into a new order of creation. . . . To quote some remarks made recently in my adopted country, Ceylon, by a Nobel Laureate in Physics, Professor C. F. Powell, "It seems to me that the evidence from astronomy and particle physics which I have described makes it possible that we are on the threshold of great and far-reaching discoveries. I have spoken of processes which, mass for mass, would bε at least a thousand times more productive of energy than nuclear energy . . . it seems that there are prodigious sources of energy in the interior regions of some galaxies, and possibly in the 'quasars,' far greater than those produced by the carbon cycle occurring in the stars. . . . And we may one day learn how to employ them."

And if Professor Powell's surmise is correct, others may already have learned, on older worlds than ours. So it would be foolish indeed to assert that the stars must be forever beyond our reach.

More than half a century ago, the great Russian pioneer space scientist Tsiolkovski wrote these moving and prophetic words: "The Earth is the cradle of the mind—but you cannot live in the cradle forever." Now, as we enter the second decade of the Age of Space, we can look still further into the future.

The Earth is indeed our cradle, which we are about to leave.

And the Solar System will be our kindergarten.

The Planets Are Not Enough

8

Altogether apart from its scientific value, space travel has one justification that transcends all others. It is probably the only way we can hope to answer one of the supreme questions of philosophy: Is Man alone in the Universe? It seems incredible that ours should be the only inhabited planet among the millions of worlds that must exist among the stars, but we cannot solve this problem by speculating about it. If it can be solved at all, it will be by visiting other planets to see for ourselves.

The Solar System, comprising the nine known worlds of our Sun and their numerous satellites, is a relatively compact structure, a snug little celestial oasis in an endless desert. It is true that millions of miles separate Earth from its neighbors, but such distances are cosmically trivial. They will even be trivial in terms of human engineering before another hundred years—a mere moment in historical time—have elapsed. However, the distances that sunder us from the possible worlds of other stars are of a totally different order of magnitude, and there are fundamental reasons for thinking that nothing—no scientific discovery or technical achievement—will ever make *them* trivial.

When today's chemical fuels have been developed to the ultimate, and such tricks as refueling in space have been fully exploited, we will have spaceships which can attain speeds of about ten miles a second. That means that the Moon will be reached in two or three days and the nearer planets in about half a year. (I am deliberately rounding these numbers off, and anyone who tries to check my arithmetic had better remember that spaceships will never travel in straight lines or at uniform speeds.) The remoter planets, such as Jupiter and Saturn, could be reached only after many years of travel, and so the trio Moon-Mars-Venus marks the practical limit of exploration for chemically propelled spaceships. Even for these cases, it is all too easy to demonstrate that hundreds of tons of fuel would be needed for each ton of payload that would make the round trip.

This situation, which used to depress the pre-atomic-energy astronauts, will not last for long. Since we are not concerned here with engineering details, we can take it for granted that eventually nuclear power, in some form or other, will be harnessed for the purposes of space flight. With energies a millionfold greater than those available from chemical fuels, speeds of hundreds, and ultimately thousands, of miles a second will be attainable. Against such speeds, the Solar System will shrink until the inner planets are no more than a few hours apart, and even Pluto will be only a week or two from Earth. Moreover, there should be no reasonable limit to the amount of equipment and material that could be taken on an interplanetary expedition. Anyone who doubts this may ponder the fact that the energy released by a single H-bomb is sufficient to carry about a million tons to Mars. It is true that we cannot as yet tap even a fraction of that energy for such a purpose, but there are already hints of how this may be done.

The short-lived Uranium Age will see the dawn of space flight; the succeeding era of fusion power will witness its fulfillment. But even when we can travel among the planets as freely as we now travel over this Earth, it seems that we will be no nearer to solving the problem of man's place in the Universe. That is a secret that will still lie hidden in the stars.

All the evidence indicates that we are alone in the Solar System. True, there is almost certainly some kind of life on Mars, and possibly on Venus—perhaps even on the Moon. (The slight evidence for lunar vegetation comes from the amateur observers who actually *look* at the Moon, and is regarded skeptically by professional astronomers, who could hardly care less about a small slag heap little more than a light-second away.) Vegetation, however, can provide little intellectual companionship. Mars may be a paradise for the botanist, but it may have little to interest the zoologist—and nothing at all to lure the anthropologist and his colleagues across some scores of millions of miles of space.

This is likely to disappoint a great many people and to take much of the zest out of space travel. Yet it would be unreasonable to expect anything else; the planets have been in existence for several billion years, and only during the last .0001 per cent of that time has the human race been slightly civilized. Even if Mars and Venus have been (or will be) suitable for higher forms of life, the chances are wildly against our encountering beings anywhere near our cultural or intellectual level at this particular moment of time. If rational creatures exist on the planets, they will be millions of years ahead of us in development—or millions of years behind us. We may expect to meet apes or angels, but never men.

The angels can probably be ruled out at once. If they existed, then surely they would already have come here to have a look at us. Some people, of course, think that this is just what they are doing. I can only say that they are going about it in a very odd manner.

We had better assume, therefore, that neither on Mars nor Venus, nor on any other of the planets, will explorers from Earth encounter intelligent life. We are the only castaways upon the tiny raft of the Solar System, as it drifts forever along the Gulf Streams of the Galaxy.

This, then, is the challenge that sooner or later the human spirit must face, when the planets have been conquered and all their secrets brought home to Earth. The nearest of the stars is a million times farther away than the closest of the planets. The spaceships we may expect to see a generation from now would take about a hundred thousand years to reach Proxima Centauri, our nearest stellar neighbor. Even the hypothetical nuclear-powered spaceships which a full century of atomic engineering may produce could hardly make the journey in less than a thousand years.

The expressive term "God's quarantine regulations" has been used to describe this state of affairs. At first sight, it appears that they are rigorously enforced. There may be millions of inhabited worlds circling other suns, harboring beings who to us would seem godlike, with civilizations and cultures beyond our wildest dreams. But we shall never meet them, and they for their part will never know of our existence.

So run the conclusions of most astronomers, even those who are quite convinced that mere common or garden interplanetary flight is just around the corner. But it is always dangerous to make negative predictions, and though the difficulties of *interstellar* travel are stupendous, they are not insuperable. It is by no means certain that man must remain trapped in the Solar System for eternity, never to know if he is a lonely freak of no cosmic significance.

There are two ways in which we might gain direct knowledge of other stellar systems without ever leaving our own. Rather surprisingly, it can be shown that radio communication would be perfectly feasible across interstellar space, if very slow-speed telegraphy were employed. However, we can hardly assume that anyone would be listening in at the precise frequency with a receiver tuned to the extremely narrow band that would have to be employed. And even if they were, it would be extremely tedious learning to talk to them with no initial knowledge of their language—and having to wait many years for any acknowledgment of our own signals, as the radio waves came limping back across the light-years. If we sent a question to Proxima Centauri, it would be almost nine years before any answer could reach Earth.

A more practical, though at first sight more startling, solution would be to send a survey ship—unmanned. This would be a gigantic extrapolation of existing techniques, but it would not involve anything fundamentally new. Imagine an automatic vessel, crammed with every type of recording instrument and controlled by an electronic brain with preset instructions. It would be launched out across space, aimed at a target it might not reach for a thousand years. But at last one of the stars ahead would begin to dominate the sky, and a century or so later, it would have grown into a sun, perhaps with planets circling around it. Sleeping instruments would wake, the tiny ship would check its speed, and its sense organs would start to record their impressions. It would circle world after world, following a program set up to cover all possible contingencies by men who had died a thousand years before. Then, with the priceless knowledge it had gained, it would begin the long voyage home.

This type of proxy exploration of the universe would be slow and uncertain and would demand long-range planning beyond the capacity of our age. Yet if there is no other way of contacting the stars, this is how it might be done. One millennium would make the investment in technical skill so that the next would reap the benefit. It would be as if Archimedes were to start a research project which could produce no results before the time of Einstein.

If men, and not merely their machines, are ever to reach the planets of other suns, problems of much greater difficulty will have to be solved. Stated in its simplest form, the question is this: How can men survive a journey which may last for several thousand years? It is rather surprising to find that there are at least five different answers which must be regarded as theoretical possibilities—however far they may be beyond the scope of today's science.

Medicine may provide two rather obvious solutions. There appears to be no fundamental reason why men should die when they do. It is certainly not a matter of the body "wearing out" in the sense that an inanimate piece of machinery does, for in the course of a single year almost the entire fabric of the body is replaced by new material. When we have discovered the details of this process, it may be possible to extend the life span indefinitely if so desired. Whether a crew of immortals, however well balanced and psychologically adjusted, could tolerate each other's company for several centuries in rather cramped quarters is an interesting subject for speculation.

Perhaps a better answer is that suggested by the story of Rip Van Winkle. Suspended animation (or, more accurately, a drastic slowing down of the body's metabolism) for periods of a few hours is now, of course, a medical commonplace. It requires no great stretch of the imagination to suppose that, with the aid of low temperatures and drugs, men may be able to hibernate for virtually unlimited periods. We can picture an automatic ship with its oblivious crew making the long journey across the interstellar night until, when a new sun was looming up, the signal was sent out to trigger the mechanisms which would revive the sleepers. When their survey was completed, they would head back to Earth and slumber again until the time came to awake once more, and to greet a world which would regard them as survivors from the distant past.

The third solution was, to the best of my knowledge, suggested over thirty years ago by Professor J. D. Bernal in a long out-of-print essay, *The World, the Flesh, and the Devil,* which must rank as one of the most outstanding feats of scientific imagination in literature. Even today, many of the ideas propounded in this little book have never been fully developed, either in or out of science fiction. (Any requests from fellow authors to borrow my copy will be flatly ignored.)

Bernal imagined entire societies launched across space, in gigantic arks which would be closed, ecologically balanced systems. They would, in fact, be miniature planets, upon which generations of men would live and die so that one day their remote descendants would return to Earth with the record of their celestial Odyssey.

The engineering, biological and sociological problems involved in such an enterprise would be of fascinating complexity. The artificial planets (at least several miles in diameter) would have to be completely self-contained and self-supporting, and no material of any kind could be wasted. Commenting on the implications of such closed systems, *Time* magazine's able, erudite science editor Jonathan Leonard once hinted that cannibalism would be compulsory among interstellar travelers. This would be a matter of definition; we crew members of the two-billion-man spaceship Earth do not consider ourselves cannibals despite the fact that every one of us must have absorbed atoms which once formed part of Caesar and Socrates, Shakespeare and Solomon.

One cannot help feeling that the interstellar ark on its thousand-year voyages would be a cumbersome way of solving the problem, even if all the social and psychological difficulties could be overcome. (Would the fiftieth generation still share the aspirations of their Pilgrim Fathers who set out from Earth so long ago?) There are, however, more sophisticated ways of getting men to the stars than the crude, brute-force methods outlined above. After the hard-headed engineering of the last few paragraphs, what follows may appear to verge upon fantasy. It involves, in the most fundamental sense of the word, the storage of human beings. And by that I do not mean anything as naïve as suspended animation.

A few months ago, in an Australian laboratory, I was watching what appeared to be perfectly normal spermatozoa wriggling across the microscope field. They *were* perfectly normal, but their history was not. For three years, they had been utterly immobile in a deep freeze, and there seemed little doubt that they could be kept fertile for centuries by the same technique. What was still more surprising, there had been enough successes with the far larger and more delicate ova to indicate that they too might survive the same treatment. If this proves to be the case, reproduction will eventually become independent of time.

The social implications of this make anything in *Brave New World* seem like child's play, but I am not concerned here with the interesting results which might have been obtained by, for example, uniting the genes of Cleopatra and Newton, had this technique been available earlier in history. (When such experiments are started, however, it would be as well to remember Shaw's famous rejection of a similar proposal: "But suppose, my dear, it turns out to have my beauty and your brains?") *

---

* We have Shaw's word for it that the would-be geneticist was a complete stranger and not, as frequently stated, Isadora Duncan.

The cumbersome interstellar ark, with its generations of travelers doomed to spend their entire lives in empty space, was merely a device to carry germ cells, knowledge, and culture from one sun to another. How much more efficient to send only the cells, to fertilize them automatically some twenty years before the voyage was due to end, to carry the embryos through to birth by techniques already foreshadowed in today's biology labs, and to bring up the babies under the tutelage of cybernetic nurses who would teach them their inheritance and their destiny when they were capable of understanding it.

These children, knowing no parents, or indeed anyone of a different age from themselves, would grow up in the strange artificial world of their speeding ship, reaching maturity in time to explore the planets ahead of them—perhaps to be the ambassadors of humanity among alien races, or perhaps to find, too late, that there was no home for them there. If their mission succeeded, it would be their duty (or that of their descendants, if the first generation could not complete the task) to see that the knowledge they had gained was someday carried back to Earth.

Would any society be morally justified, we may well ask, in planning so onerous and uncertain a future for its unborn—indeed unconceived—children? That is a question which different ages may answer in different ways. What to one era would seem a cold-blooded sacrifice might to another appear a great and glorious adventure. There are complex problems here which cannot be settled by instinctive, emotional answers.

So far, we have assumed that all interstellar voyages must of necessity last for many hundreds or even thousands of years. The nearest star is more than four light-years away; the Galaxy itself—the island Universe of which our Sun is one insignificant member—is hundreds of thousands of light-years across; and the distances *between* the galaxies are of the order of a million light-years. The speed of light appears to be a fundamental limit to velocity; in this sense it is quite different from the now outmoded "sound barrier," which is merely an attribute of the particular gases which happen to constitute our atmosphere.

Even if we could reach the speed of light, therefore, interstellar journeys would still require many years of travel, and only in the case of the very nearest stars would it appear possible for a voyager to make the round trip in a single lifetime, without resort to such techniques as suspended animation. However, as we shall see, the actual situation is a good deal more complex than this.

First of all, is it even theoretically possible to build spaceships capable of approaching the speed of light? (That is, 186,000 miles a second or 670,000,000 miles per hour.) The problem is that of finding a sufficient source of energy and applying it. Einstein's famous equation $E = mc^2$ gives an answer—on paper—which a few centuries of technology may be able to realize in terms of engineering. If we can achieve the *total* annihilation of matter—not the conversion of a mere fraction of a per cent of it into energy—we can approach as near to the speed of light as we please. We can never reach it, but a journey at 99.9 per cent of the speed of light would, after all, take very little longer than one at exactly the speed of light, so the difference would hardly seem of practical importance.

Complete annihilation of matter is still as much a dream as atomic energy itself was thirty years ago. However, the discovery of the anti-proton (which engages in mutual suicide on meeting a normal proton) may be the first step on the road to its realization.

Traveling at speeds approaching that of light, however, involves us at once in one of the most baffling paradoxes which spring from the theory of relativity—the so-called "time-dilation effect." It is impossible to explain *why* this effect occurs without delving into very elementary yet extremely subtle mathematics. (There is nothing difficult about basic relativity math: most of it is simple algebra. The difficulty lies in the underlying concepts.) Nevertheless, even if the explanation must be skipped, the results of the time-dilation effect can be stated readily enough in nontechnical language.

Time itself is a variable quantity; the rate at which it flows depends upon the speed of the observer. The difference is infinitesimal at the velocities of everyday life, and even at the velocities of normal astronomical bodies. It is all-important as we approach to within a few per cent of the speed of light. To put it crudely, the faster one travels, the more slowly time will pass. At the speed of light, time would cease to exist; the moment "Now" would last forever.

Let us take an extreme example to show what this implies. If a spaceship left Earth for Proxima Centauri at the speed of light, and came back at once at the same velocity, it would have been gone for some eight and one-half years according to all the clocks and calendars of Earth. *But the people in the ship, and all their clocks, would have recorded no lapsed time at all.*

At a physically attainable speed, say 95 per cent of the velocity of light, the inhabitants of the ship would think that the round trip had lasted about three years. At 99 per cent, it would have seemed little more than a year to them. In each case, however, they would return more than eight years—Earth time—after they had departed. (No allowance has been made here for stopping and starting, which would require additional time.)

If we imagine a more extensive trip, we get still more surprising results. The travelers might be gone for a thousand years, from the point of view of Earth, having set out for a star five hundred light-years away. If their ship had averaged 99.9 per cent of the speed of light, they would be fifty years older when they returned to an Earth—*where ten centuries had passed away!* *

It should be emphasized that this effect, incredible though it appears to be, is one of the natural consequences of Einstein's theory. The equation connecting mass and energy once appeared to be equally fantastic and remote from any practical application. It would be very unwise, therefore, to assume that the equation linking time and velocity will never be of more than theoretical interest. Anything which does not violate natural laws must be considered a possibility—and the events of the last few decades have shown

---

* The physical reality of the time-dilation effect has been the subject of unusually acrimonious debate in recent years. Very few scientists now have any doubt of its existence, but its magnitude may not have the values quoted above. My figures are based on special relativity, which is too unsophisticated to deal with the complexities of an actual flight.

clearly enough that things which are possible will always be achieved if the incentive is sufficiently great.

Whether the incentive will be sufficient here is a question which only the future can answer. The men of five hundred or a thousand years from now will have motivations very different from ours, but if they are men at all they will still burn with that restless curiosity which has driven us over this world and which is about to take us into space. Sooner or later we will come to the edge of the Solar System and will be looking out across the ultimate abyss. Then we must choose whether we reach the stars—or whether we wait until the stars reach us.

When the Aliens Come

9

The first encounter between Earthman and Alien is one of the oldest and most hackneyed themes of science-fiction. Indeed, it has now become such a cliché that "take me to your leader" jokes are perfectly familiar even to those benighted souls who have never read a word of s.f. in their lives.

How odd, therefore, that there seem to be so few serious *factual* discussions of this subject. True, there have been essays without number on the possibilities of extraterrestrial life and the ways we might communicate with it, but most of them stop abruptly at the really interesting point. The astronomers and biologists, and even the philosophers and theologians, have all had their say in the last few years. But the sociologists and politicians have left it to the science-fiction writers—and this at the very time when the subject is moving out of the realm of fantasy.

All War Departments, it is said (though one sometimes doubts this), have plans worked out for every conceivable eventuality. Presumably somewhere in the Pentagon are the orders for such lamentable necessities as the invasion of Canada or the bombing of London—or even New York, *vide Fail Safe*. If there are any plans for the defense of Earth, no one has ever mentioned them.

Probably the Department of Defense would argue, if pressed, that the matter was under the jurisdiction of the State Department— and you may be quite surprised to learn that State *does* have an

93

"Office of Outer Space Affairs." On March 15, 1967, its director, Robert F. Packard, presented a paper on "The Role of the Diplomat" to the Fifth Goddard Memorial Symposium in Washington. It was concerned, however, exclusively with terrestrial diplomats and did not even hint that there might be any other kind. In the absence of any official guidance, therefore, let us attempt to construct some scenarios (I believe this is the approved term among the nuclear Doomsday planners) of our own.

The first problem we have to face is our total ignorance of the nature of extraterrestrials (ETs)—we do not even know if they exist! If they don't, of course, that is the end of the matter—but even if this is true, *we can never be sure.* And the idea that *we* are the only intelligent creatures in a cosmos of a hundred million galaxies is so preposterous that there are very few astronomers today who would take it seriously.

It is safest to assume, therefore, that They *are* out there and to consider the manner in which this fact may impinge upon human society. It could come in ways as undramatic as the deciphering of an ancient papyrus, or as shattering as a crash landing, with ray guns ablaze, on the White House lawn.

The most probable scenario, at least during the foreseeable future, might be called "Discovery Without Contact." By this I mean that we obtain unequivocal proof that intelligent ETs exist (or have existed), but in a manner that excludes communication.

Such a proof might be obtained from archaeology or geology. The discovery of a fossilized transistor radio in an undisturbed coal bed, preferably accompanied by skeletons that did not fit into any evolutionary tree, would be convincing evidence that our planet was once visited from space. Ancient legends, wall paintings, or other works of art might also record such visits in historic times; unfortunately, this type of evidence can only be circumstantial—it can never be conclusive.

Shklovskii and Sagan's fascinating book *Intelligent Life in the Universe* reproduces some three-thousand-year-old Babylonian seals which, together with their associated legends, can very easily be taken to depict encounters between men and non-men; parts of the Bible have been interpreted in the same manner. However, the mythmaking abilities of the human mind are so unlimited that it would be very foolish to accept these items as proof of anything. After all, what would intelligent aliens make of a Superman comic strip?

No; in a matter as important as this, the only acceptable evidence would be hardware. About twenty years ago, in a short story "The Sentinel" (which Stanley Kubrick later used as the basis of *2001: A Space Odyssey*), I suggested that the best place to look for such evidence would be on a relatively stable and changeless world like the Moon. On Earth, with its incessant weather and geological upheavals, no extraterrestrial artifact would survive for very long, though this is no excuse for not keeping our eyes open. The reason why space hardware has never been discovered may simply be because no archaeologist ever dreamed of looking for it.

Although the philosophical—and sensational—impact of such a discovery would be enormous, after the initial excitement had ebbed the world would probably continue on its way much as before. Once he had read a few Sunday supplements and watched a few TV specials, the proverbial man in the street would say: "This is all very interesting, but it happened a long time ago and hasn't anything to do with me. Sure, They could come back one day, but there are plenty of more important things to worry about." And he would be quite right.

Almost every field of scientific inquiry, however, would be profoundly affected. If it appeared that the visitors came from one of the other worlds of our own Solar System—Mars, for example— this would obviously be a great stimulus to planetary exploration, but it would also start us searching much farther afield.

Two intelligent races in the same Solar System, even if they were separated by millions of years of time, would provide virtually conclusive proof that higher civilizations were very common throughout the Universe. This would immediately stimulate really determined attempts to detect signals from other star systems.

Little more than a decade ago, the astronomers suddenly realized, to their considerable surprise, that our radio technology has advanced to the point where we can start talking seriously about interstellar communication. And if, after only fifty years, *we* have reached such a level of development, what might older civilizations have achieved?

Scattered among the stars there may be radio beacons and transmitters of unimaginable power; the British cosmologist Fred Hoyle has expressed the view that there may be a kind of Galactic communications network, linking thousands or millions of worlds. Within a very few centuries, we may be clever enough to plug ourselves into the circuit; it may take us a little longer to understand what the other subscribers are saying. (Conceivably, "Get off the line!")

The possibilities opened up even by one-way communication (passive eavesdropping) are almost unlimited. The signals would certainly contain visual material—not necessarily real-time TV—which it would be rather easy to reconstruct. And then, across the light-years, we would be able to look at other worlds and other races. . . .

Now this is a situation far more exciting than the discovery of fossil artifacts. We would be dealing not with prehistory, but with *news*—though news that had been slightly delayed in transit. If the signals came from the very closest stars, they would have left their transmitters only five or ten years ago; a more likely time lag would be a few centuries. In any event, we would be listening to civilizations still in existence, not studying the relics of vanished cultures.

The things we could learn might change our own society beyond recognition. It would be as if the America of Lincoln's time could tune into the TV programs of today; though there would be much that could not be understood, there would also be clues that could leapfrog whole technologies into the future. (Ironically enough, the commercials would contain some of the most valuable information!) Nineteenth-century viewers would see that heavier-than-air machines were possible, and simple observation would reveal the principles of their design. The still unimagined uses of electricity would be demonstrated (the telephone, the electric light . . .), and this would be enough to set scientists on the right track. For knowing that a thing *can* be done is more than half the battle.

As signals from the stars could be received only by nations possessing very large radio telescopes, there would be the opportunity —and the temptation—to keep them secret. Knowledge is the most precious of all commodities, and it is a strange thought that the balance of power may one day be shifted by a few micromicrowatts collected from the depths of space. Yet it should no longer surprise us; for who dreamed, fifty years ago, that the faint flicker of dying atoms in a physics lab would change the course of history?

Glimpses of supercivilizations could have either stimulating or stultifying effects on our society. If the technological gulf was not too great to be bridged, and the programs we intercepted contained hints and clues that we could understand, we would probably rise to the challenge. But if we found ourselves in the position of Neanderthalers confronted by New York City, the psychological shock could be so great that we might give up the struggle. This appears to have happened on our own world from time to time, when primitive races have come into contact with more advanced ones. We will have a good chance of studying this phenomenon in a very few years, when communications satellites start beaming *our* TV programs into such places as the Amazon jungle. This is the last century during which widely disparate cultures will exist on Earth; would-be students of astrosociology should make the most of their opportunity before it vanishes forever. And no one will be surprised to hear that Margaret Mead is intensely interested in space flight. . . .

The discovery of an active communications network in our region of space (and I would make a small bet that such a thing exists) would at once raise a very difficult problem: Should we announce our presence by joining in the conversation, or should we maintain a discreet silence? If anyone thinks that this is an easy question to answer, let him put himself in the place of a cultured and sensitive extraterrestrial whose knowledge of human civilization is based largely on "The Man from UNCLE," "Dragnet," and "The Late, Late Show."

Probably everyone would agree that the wisest plan would be to listen carefully until we had learned as much as possible, before attempting to signal our presence. However, such caution may already be much too late; as far as Earth is concerned, the electronic cat was let out of the bag a couple of decades ago. Although it is unlikely that our first radio programs have ever been monitored (they were too low-powered, and at unfavorable frequencies), the megawatt radars developed during the Second World War may have been detected tens of light-years away. We have been making such a din that the neighbors can hardly have overlooked us, and I sometimes wonder when they will start banging on the walls.

Of course, if intelligent civilizations are so far apart that no *physical* transport between them is possible (as most scientists believe), then there would seem no objection to announcing our presence. As the old jingle puts it, "Sticks and stones can break my bones, but words will never hurt me." Some writers have argued that we should be thankful for the immense distances of interstellar space. Cosmic communities can talk to one another for their mutual benefit, but they can never do each other any harm.

However, this is a naïve and unrealistic view. Even if star travel is impossible (later we will give reasons for believing that, on the contrary, it is rather easy), "mere" communications could do a lot of damage. After all, this is the basis on which all censors act. A really malevolent society could destroy another one quite effectively by a few items of well-chosen information. ("Now, kiddies, after you've prepared your uranium hexafluoride . . .")

In any case, after a certain level of technical sophistication it is meaningless to distinguish between the transfer of material objects and the transfer of information. Fred Hoyle, in his novel *A for Andromeda,* has suggested that a sufficiently complex signal from space might serve as the genetic blueprint for constructing an extraterrestrial entity. An invasion by radio may seem a little farfetched, but it does not involve any scientific impossibilities.

I suspect that once we had heard voices echoing between the stars, it would not be long before curiosity—or egotism—made us join the conversation. However, the task of framing suitable replies might be difficult. Naturally, we would present ourselves in the best possible light, and the temptation to gloss over unflattering aspects of human history and behavior would be considerable. Also—who would speak for man? It is easy to imagine our current ideologies proclaiming their rival merits to the heavens, and even a super-civilization might well be baffled by propaganda blasts based on the teachings of Chairman Mao.

Perhaps fortunately, the power and the resources needed to beam a profile of *Homo sapiens* across interstellar space are so great that a global, cooperative effort would be needed. Then, for the first time, mankind might speak with a single voice; and the problem of compiling the program might induce a certain humility.

After that, there would come the long wait for the answer. In the unlikely event that there is a civilization circling the very nearest star—Proxima Centauri—we could not receive a reply in less than eight years. It is more probable that the delay would be measured in decades, so any two-way conversations would be distinctly tedious. They would, in fact, be long-term research projects, with scientists receiving in their old age answers to questions they had asked in their youth.

Despite its unavoidable slowness, such conversation-without-contact would, over the centuries, have enormous and perhaps decisive effects upon human society. Quite apart from the technological leapfrogging already mentioned, it could produce knowledge of different races, patterns of thought, and political systems that would completely change our philosophical and religious views. Are good and evil man-made concepts? Do other races have gods, and of what nature? Is death universal? These are a few of the questions we might ask of the stars, and some of the answers might not be to our liking.

Yet perhaps the most important result of such contacts might be the simple proof that other intelligent races do exist. Even if our cosmic conversations never rise above the "Me Tarzan—You Jane" level, we would no longer feel so alone in an apparently hostile universe. And, above all, knowledge that other beings had safely passed their nuclear crises would give us renewed hope for our own future. It would help to dispel present nagging doubts about the survival value of intelligence. We have, as yet, no definite proof that too much brain, like too much armor, is not one of those unfortunate evolutionary accidents that leads to the annihilation of its possessors.

If, however, this dangerous gift can be turned to advantage, then all over the Universe there must be races that have been gathering knowledge, and perfecting their technologies, for periods of time that may be measured in millions of years. Anything that is theoretically possible, and is worth doing, will have been achieved. Among those achievements will be the crossing of interstellar space.

Travel to the stars requires no more energy and demands no more of propulsion systems than flight to the nearest planets. There are rockets in existence today that could launch tonnage payloads to Proxima Centauri; however, it would take them about a quarter of a million years to get there—and Proxima, remember, is the very closest of our stellar neighbors. We will have to move a little faster.

But even at the speed of light (about twenty thousand times greater than that of any space probe yet built) Proxima is still four years away, and it would take over a hundred thousand years to cross the width of the Galaxy.

Yet this does not prove, as many scientists have rashly argued, that interstellar flight is impossible. There are several ways in which it might be achieved, by technologies which even we can imagine, and which might be within our grasp a few centuries from today.

It is highly probable—though not absolutely certain—that the velocity of light can never be exceeded by any material object.* Star travel will thus be very time-consuming; the duration of voyages will be measured in decades at the very least—more likely, in mil-

* See "Possible, That's All!," page 108.

lennia. For such short-lived creatures as human beings, this would require multigeneration trips in totally enclosed, self-contained mobile worldlets (little Earths)—or, perhaps less technically demanding, some form of suspended animation.

There is another factor which is almost invariably overlooked in discussions of star travel. Our understandable doubts about the practicability and desirability of such ventures would not be shared by really advanced creatures, who might have unlimited life spans. If we were immortal, the stars would not seem very far away.

It is, therefore, quite unrealistic *not* to expect visitors from deep space, sooner or later. And, of course, a great many people—not all of them cranks—think that they are arriving right now.

UFOlogy is a can of worms into which I refuse to probe.* Let us take the line of least resistance and assume that the strange apparitions whizzing through our skies are indeed of extraterrestrial origin, and that this is finally proved beyond all reasonable doubt.

The first result would be a drastic lowering of the international temperature; any current wars would rapidly liquidate themselves. This point has been made by numerous writers—starting with André Maurois, whose *War Against the Moon* suggested half a century ago that the only way to secure peace on Earth would be to manufacture a fake menace from space. A genuine one would be even more effective.

If, however, the ETs did nothing but merely study us like detached anthropologists, eventually we would resume our pastimes—including minor wars—though with a certain tendency to keep looking over our shoulders. Anyone who has observed the neat farms on the slopes of a volcano will agree that the human race has an astonishing ability to continue life as if nothing has happened even when something very obviously has. We can be sure, though, that under the cover of normalcy there would be heroic attempts by all the secret services and intelligence agencies to establish contact with the aliens—for the exclusive benefit of their respective coun-

* But see "Things in the Sky," page 203.

tries. Every astronomical observatory in the Free World would be pelted with largesse from the CIA.

Such a situation, though it might endure for a decade or so, could not be stable. Sooner or later there would be a communications breakthrough, or else the human race would become so exasperated by the spectacle of Olympian indifference that an "Aliens Go Home!" movement would develop. Rude radio noises would eventually escalate to nuclear bombs, at which point the aliens either *would* go home or would take steps to abate the nuisance.

It has often been suggested that the arrival of visitors from space would cause widespread panic; for this reason, some UFO enthusiasts believe that the U.S. government is keeping the "facts" concealed. (Actually, the reverse is nearer the truth; as one Pentagonian once remarked sourly, "If there really *were* flying saucers, all us majors would be colonels.") The world has become much more sophisticated since the far-off days of Orson Welles's famous radio broadcast. It is unlikely that a friendly or neutral contact— except in primitive communities, or by creatures of outrageous appearance—would produce an outburst of hysteria like that which afflicted New Jersey in 1938. Thousands of people would probably rush to their cars, but they would be in a hurry to get *to* the scene of such an historic event, not to escape from it.

And yet, having written those words, I begin to wonder. It is easy to be calm and collected when discussing a theoretical possibility; in the event, one's behavior may be very different. Like any reasonably observant person who lives under clear skies, I have seen a good many objects that could have been taken for UFOs, and on just one occasion, it seemed as if it might be the "real thing." (No one will ever believe this, but I was with Stanley Kubrick, the very night we decided to make our movie.) * I shall never forget the feelings of awe and wonder—yes, and fear—that chased each other through my mind, before I discovered that the object was only Echo 1, seen under somewhat unusual conditions.

No one can be sure how *he* would react in the presence of a vis-

* See "Son of Dr. Strangelove," page 240.

itor from another world. When the time comes to announce that mankind is no longer alone, those who prepare and issue the statement will have a truly terrifying responsibility. Though they will certainly try to sound reassuring, they will know that they are whistling in the dark.

It is impossible to guess at all the motivations which might drive ETs to visit our planet. Human societies have an almost unbelievable range of behavior, and totally alien cultures might act in ways quite incomprehensible to us. Anyone who doubts this should attempt to look at our own society from outside, and imagine himself in the role of an intelligent Martian trying to understand what was really going on at a political rally, a chess tournament, the floor of the Stock Exchange, a religious revival, a symphony concert, a baseball game, a sit-in, a TV quiz program—the list is endless.

In a witty essay on "Extraterrestrial Linguistics" Professor Solomon Golumb of the University of Southern California has tried to make order out of chaos by suggesting that our neighbors might wish to deal with us under one or another of these headings: (1) Help! (2) Buy! (3) Convert! (4) Vacate! (5) Negotiate! (6) Work! (7) Discuss! And a famous short story of Damon Knight's has added (8) Serve! (Broiled or fried.)

Yet even this rather comprehensive list assumes that They possess psychologies similar to ours, and that we can make mental or at least physical contact. Some ingenious science-fiction writers have argued that this may not necessarily be the case. In Olaf Stapledon's tremendous history of the future, *Last and First Men,* the Earth was invaded by microscopic creatures from Mars who formed a rational entity only when they coalesced into a kind of intelligent cloud. (If this seems far-fetched, consider how many independently viable living cells go to form the entity you are pleased to call You.) Because Stapledon's Martians found it very exhausting to assume the solid state, they worshipped hard, rigid bodies and thus avidly collected diamonds and other gems, while ignoring the soft, semiliquid creatures who transported these sacred objects. They were aware of automobiles, but not of human beings. . . .

Indeed, it has been suggested that any dispassionate observer of the United States would conclude that the automobile was its dominant life form.

It would be difficult to bridge such a psychophysical gulf; a similar one may already exist right here on Earth, between man and such social insects as ants, termites, or bees. Here the individual is nothing: the state is all, beyond the wildest dreams of any totalitarian dictator.

In extreme cases, we might not even be able to *detect* an alien species, except by rather sophisticated instruments. It could be gaseous, or electronic, or could operate on time-scales hundreds of times faster or slower than ours. Even human beings live at different rates, judging by speeds of conversation, and there seems little doubt that dolphins think and speak much more rapidly than we do, though they are courteous enough to use slow-speed baby talk when we attempt to communicate with them.

I mention these rather far-out speculations not because I take them very seriously (I don't) but because they show the utter lack of imagination of those who think that intelligent aliens must be humanoid. Now there may well be millions of intelligent humanoid races in the Universe, since ours appears to be a successful and practical design. But even if all the ingredients are exactly the same, and in approximately the same places, it would be exceedingly rare to find a humanoid alien who resembled a man as closely as does, say, a chimpanzee.

I would even go so far as to say that, from the cosmic viewpoint, *all* terrestrial mammals are "humanoid." They all have four limbs, two eyes, two ears, one mouth, arranged symmetrically about a single axis. Could a visitor from Sirius really tell the difference between a man and, for example, a bear? ("I'm terribly sorry, Mr. Prime Minister, but *all* humanoids look the same to me. . . .")

Even if we restrict ourselves to the sense organs, and manipulators, with which we are familiar on Earth, they could be arranged— and, equally important, used—in an enormous variety of ways, to produce effects of astonishing strangeness. The late Nobel Laureate

Dr. Hermann Muller expressed this very well in his phrase "the bizarreness of the right and proper." An alien, he pointed out, "would find it most remarkable that we had an organ combining the requirements of breathing, ingesting, chewing, biting, and on occasion fighting, helping to thread needles, yelling, whistling, lecturing, and grimacing. He might well have separate organs for all these purposes, located in diverse parts of his body, and would consider as awkward and primitive our imperfect separation of these functions."

Even judging by the examples on our own world, where all life is based on the same biochemical system, the ingenuity of nature seems almost unlimited. Consider the nightmare shapes of the deep sea or the armored gargoyles of the insect world; we may one day encounter rational creatures in forms analogous to all of these. And, conversely, we should not be misled by superficial resemblances; think of the abyss that separates the sharks from their almost-duplicates, the dolphins. Or, nearer home, that which tragically divides the sundered children of Abraham today. . . .

So beyond doubt, physical shape is unimportant compared with motivation. Once again, because of our blinkered human viewpoint, we cannot extend our ideas much beyond Dr. Golumb's not altogether facetious list of directives. Now although everything that is conceivable will occur at least once, in our Galaxy of a hundred billion suns, some of these categories seem more likely than others. The insanely malevolent invaders beloved by the horror comics have perhaps the least plausibility—if only because they would have destroyed themselves long before they got to us. Any race intelligent enough to conquer interstellar space must first have conquered its own inner demons.

Moreover, there seem few grounds for cosmic conflict, even if it were technically possible. It is hard to see what attractions our world could offer visitors from space; since their physical forms and requirements would be totally different from ours, it is very unlikely that they would be able to live here.

There are no material objects—no conceivable treasures or

spices or jewels or exotic drugs—valuable enough to justify the conquest of a world. Anything we possess, *they* could manufacture easily enough at home. For imagine what *our* chemists will have done, a thousand years from now.

There may, of course, be entities who collect Solar Systems as a child may collect stamps. If this happened to us, we might never be aware of it. What do the inhabitants of a beehive know of their keeper?

That may be an analogy worth pursuing. Men do not interfere with bees—or wasps—unless they have good reasons: as far as possible, they prefer to leave them alone. Though we possess no better weapons than 100-megaton bombs, we are not entirely defenseless, and even an advanced supercivilization might think twice about tangling with us.

If they were desperate—if, for example, they were the last survivors of an ancient race, whose mobile worldlet had almost exhausted its supplies after aeons of voyaging—they might be tempted to make a fresh home in our Solar System. But even in that case, cooperation would be to their advantage—and to ours. Since they would probably be able to transmute any element into any other, there is no reason why they should covet Earth. The barren Moon and the drifting slag heaps of the asteroid belt would provide all the raw materials they needed—and the Sun, all the energy. Our planet intercepts only one part in two billion of the radiation pouring from the Sun, and we actually utilize only a minute fraction of that. There is matter and energy enough in the Solar System for many civilizations, for ages to come.

Unfortunately, our record so far has not shown much inclination for peaceful coexistence. If such writers as Robert Ardrey are correct, much of human (and animal) behavior is determined by the concept of "territoriality." The landowner who places a sign on a piece of private wilderness announcing that "Trespassers Will Be Prosecuted" would then speak for his entire species. If some inoffensive visitors began to colonize the frozen outer moon of Jupiter, there would be angry voices proclaiming it sacred soil, and retired generals would warn us to keep our lasers dry, and not to fire until we can see the greens of their eyes.

All of which leads to a conclusion that may not be very original but whose importance cannot be overstressed. Everyone recognizes that our present racial, political, and international troubles are symptoms of a sickness which must be cured before we can survive on our own planet—but the stakes may be even greater than that.

Though it is impossible to guard against all the eventualities that the future may bring, if we can learn to live with ourselves we will at least improve our chances of living with aliens. And that word "ourselves" should be interpreted in the widest possible context—to embrace, as far as practical, *all* intelligent creatures on this planet. At the moment, in a paroxysm of greed and folly, we have virtually exterminated the largest animal this world has ever seen. Only a few eccentrics have felt any twinges of conscience over the fact that the brain of a blue whale is larger than a man's, so that we do not know what kind of entity we have really destroyed.

It is true that our aggressive instincts, inherited from the predatory apes who were our ancestors, have made us masters of this planet and have already propelled us into space. Without those instincts, we might have perished long ago; they have served us well. But, to quote the ruler of Camelot, "The old order changeth, giving place to new. . . . Lest one good custom should corrupt the world."

We have the intelligence to change, or at least to control, the atavistic urges programmed into our behavior. Though it may seem a paradox, and a denial of all past history, gentleness and tolerance may yet prove to have the greatest survival value, when we move out into the cosmic stage.

If this is true, let us hope that we have time to cultivate these virtues. For the hour is very late, and no one can guess how many strange eyes and minds are already turned upon the planet Earth.

Possible, That's All!

10

The Galactic novels of my esteemed friend Dr. Asimov gave me such pleasure in boyhood that it is with great reluctance that I rise up to challenge some of his recent statements ("Impossible, That's All," *Magazine of Fantasy & Science Fiction,* February 1967). I can only presume that advancing years, and the insatiable demands of the Asimov-of-the-Month-Club Selection Board, have caused a certain enfeeblement of the far-ranging imagination that has delighted so many generations of s.f. fandom. [Firm note from editor: To put the record straight, and to prevent any innocent readers' being deceived—the Good Doctor was born three years *later* than his Reluctant Critic.]

The possibility, or otherwise, of speeds greater than that of light cannot be disposed of quite as cavalierly as Dr. A. does in his article. First of all, even the restricted, or Special, Theory of Relativity does not deny the existence of such speeds. It only says that speeds *equal* to that of light are impossible—which is quite another matter.

The naïve layman who has never been exposed to quantum physics may well argue that to get from below the speed of light to above it one has to pass *through* it. But this is not necessarily the case; we might be able to jump over it, thus avoiding the mathematical disasters which the well-known Lorentz equations predict when one's velocity is *exactly* equal to that of light. Above this critical speed, the equations can scarcely be expected to apply, though if certain interesting assumptions are made, they may still do so.

I am indebted, if that is the word, to Dr. Gerald Feinberg of Columbia University for this idea. His paper "On the Possibility of Superphotic Speed Particles" points out that since sudden jumps from one state to another are characteristic of quantum systems, it might be possible to hop over the "light barrier" without going through it. If anyone thinks that this is ridiculous, I would remind him that quantum-effect devices doing similar tricks are now on the market—witness the tunnel diode. Anything that can rack up sales of hundreds of thousands of dollars should be taken very seriously indeed.

Even if there is no way through the light barrier, Dr. Feinberg suggests that there may be another universe on the other side of it, composed entirely of particles that cannot travel *slower* than the speed of light. (Anyone who can visualize just what is meant by that phrase "on the other side of" is a much better man than I am.) However, as such particles—assuming that they still obey the Lorentz equations—would possess imaginary mass or negative energy, we might never be able to detect them or use them for any practical purpose like interstellar signaling. As far as we are concerned, they might as well not exist.

This last point does not worry me unduly. Similar harsh things were once said about the neutrino, yet it is now quite easy to detect this improbable object, if you are prepared to baby-sit a few hundred tons of equipment for several months two miles down in an abandoned gold mine. Anyway, mere trifles like negative energy and imaginary mass should not deter any mathematical physicist worthy of his salt. Odder concepts are being bandied round all the time in the quark-infested precincts of Brookhaven and CERN.

Perhaps at this point I should exorcise a phantom which, rather wisely, the Good Doctor refrained from invoking. There are many things that *do* travel faster than light, but they are not exactly "things." They are only appearances, which do not involve the transfer of energy, matter, or information.

One example—familiar to thousands of radar technicians—is the movement of radio waves along the rectangular copper pipes known as wave guides. The electromagnetic patterns traveling through a wave guide can *only* move faster than light—never at less than this speed! But they cannot carry signals; the *changes* of pattern which alone can do this move more slowly than light, and by precisely the same ratio as the others exceed it. (I.e., the product of the two speeds equals the square of the speed of light.)

If this sounds complicated, let me give an example which I hope will clarify the situation. Suppose we had a wave guide one light-year long, and fed radio signals into it. Under no circumstances could anything emerge from the other end in less than a year; in fact, it might be ten years before the message arrived, moving at only a tenth of the speed of light. But once the waves had got through, they would have established a pattern that swept along the guide at ten times the speed of light. Again I must emphasize that this pattern *would carry no information*. Any signal or message would require a change of some kind at the transmitter, which would take ten years to make the one-light-year trip.

If you have ever watched storm waves hitting a breakwater you may have seen a very similar phenomenon. When the line of waves hits the obstacle at an acute angle, a veritable waterspout appears at the point of intersection and moves along the breakwater at a speed which is always greater than that of the oncoming waves and can have any value up to infinity (when the lines are parallel, and the whole sea front erupts at once). But no matter how ingenious you are, there is no way in which you could contrive to use this waterspout to carry signals—or objects—along the coast. Though it contains a lot of energy, it does not involve any *movement* of that energy. The same is true of the hyperspeed patterns in a wave guide.

If you wish to investigate this subject in more detail, I refer you to the article by s.f. old-timer Milton A. Rothman, "Things That Go Faster Than Light" in *Scientific American* for July 1960. The essential fact is that the existence of such speeds ("phase velocities") in no way invalidates the Theory of Relativity.

However, there may be other phenomena that do precisely this. Please fasten your seatbelts and read—slowly—the following extract from a letter by Professor Herbert Dingle, published in the Royal Astronomical Society's Journal, *The Observatory,* for December 1965, (*85,* 949, pp. 262–64.) It will repay careful study.

The recent report that artificially produced messages from distant parts of the universe might have been detected has prompted much speculation on the possibility of communication over long distances, in all of which it seems to have been taken for granted that a time of at least r/c must elapse before a signal can be received from a distance r (c = velocity of light). There is, however, no evidence for this. There is reason for believing that it is true for any phenomenon uniquely locatable at a point, or small region, but some phenomena are not so locatable. If the postulate of relativity (i.e., the postulate that there is no natural standard of rest, so that the relative motion of two bodies cannot be uniquely divided between them) is true, the Doppler effect affords a means of *instantaneous communication over any distance at all.* . . . A code of signals . . . could therefore be devised that, in principle, *would enable us to send a message to any distance and receive an immediate reply.* [My italics.]

Unfortunately we do not know whether the postulate of relativity is true or not . . . Since the demise of the special relativity theory has not yet succeeded in penetrating into general awareness, it seems worth while to show its impotence in the present problem. . . .

And so on, for a few hundred words of tight mathematical logic, followed by a reply from a critic which Professor Dingle demolishes, at least to his satisfaction, in the August 1966 issue of the *Observatory.* I won't give details of the discussion because they are far too technical for this journal. (Translation: I don't understand them.)

The point I wish to stress, however, should already have been made sufficiently clear by the extract. Despite its formidable success in many *local* applications, relativity may not be the last word about the Universe. Indeed, it would be quite unprecedented if it were.

The *General* Theory—which deals with gravity and accelerated motions, unlike the Special Theory, which is concerned only with unaccelerated motion—may already be in deep trouble. One of the world's leading astrophysicists (he may have changed his mind now, so I will not identify him beyond saying that his name begins with Z) once shook me by remarking casually, as we were on the way up Mt. Palomar, that he regarded all the three "proofs" of the General Theory as disproved. And only this week I read that Professor Dicke has detected a flattening of the Sun's poles, which accounts for the orbital peculiarities of Mercury, long regarded as the most convincing evidence for the theory.

If Dicke is right, the fact that Einstein's calculations gave the correct result for the precession of Mercury will be pure coincidence. And *then* we shall have an astronomical scandal at both ends of the Solar System: for Lowell's "prediction" of Pluto's orbit also seems completely fortuitous. Pluto is far too small to have produced the perturbations that led to its discovery. (Has anyone yet written a story suggesting that it is the satellite of a much larger but invisible planet?)

And now that I have got started, I'd like to take a swipe at another Einsteinian sacred cow—the Principle of Equivalence, which is the basis of the gravitational theory. Every book on the subject—George Gamow's *Gravity* is a good example—illustrates the principle by imagining a man in a spaceship. If the spaceship is accelerating at a steady rate, it is claimed that there is no way in which the occupant can distinguish the "inertial" forces acting on him from those due to a gravitational field.

Now this is patent nonsense—unless the observer and his spaceship have zero dimensions. One can *always* distinguish between a gravitational field and an inertial one. For if you examine any gravitational field with a suitable instrument (which need be no more complicated than a couple of ball bearings, whose movements in free fall are observed with sufficient precision), you will quickly discover two facts: (1) the field varies in intensity from point to point, because it obeys an inverse square law (this "gravity gradient" effect is now used to stabilize satellites in orbit); (2) the field is not parallel, as it radiates from some central gravitating body.

But the "pseudo-gravitational" force due to acceleration can—in principle, at least—be made uniform and parallel over as great a volume of space as desired. So the distinction between the two would be obvious after a very brief period of examination.

Let us ignore that unpleasant little man in the front row who has just popped up to ask how I've spotted a flaw missed by Albert Einstein and some 90 per cent of all the mathematicians who have ever lived on this planet, since the beginning of time. But *if* the Principle of Equivalence is invalid, several important consequences follow. One of the most effective arguments against the possibility of antigravitational and "space drive" devices is overthrown—surely a consummation devoutly to be wished by all advocates of planetary exploration, not to mention those billions who will shortly be cringing beneath the impact of sonic thunderbolts from SSTs. In addition, and more relevant to our present argument, we will have made a hole in the Theory of Relativity through which we should be able to fly a superphotic ship.

Talking of holes leads rather naturally to our old friend the space warp, that convenient short-cut taken by so many writers of interstellar fiction (myself included). As a firm believer in Haldane's law ("The Universe is not only queerer than we imagine; it is queerer than we *can* imagine"), I think we should not dismiss space warps merely as fictional devices. At least one mathematical physicist—Profesor J. A. Wheeler—has constructed a theory of space-time which involves what he has picturesquely called "worm-holes." These have all the classic attributes of the space warp; you disappear at A and reappear at B, without ever visiting any point in between. Unfortunately, in Wheeler's theory the average speed from A to B, even via worm-hole, still works out at less than the speed of light. This seems very unenterprising and I hope the professor does a little more homework.

Another interesting, and unusual, attempt to demolish the light barrier was made in the last chapter of the book *Islands in Space,* by Dandridge M. Cole and Donald W. Cox. They pointed out that all the tests of the relativity equations had been carried out by particles accelerated by *external* forces, not self-propelled systems like rockets. It was unwise, they argued, to assume that the same laws applied in this case.

And here I must admit to a little embarrassment. I had forgotten, until I referred to my copy, that the preface of *Islands in Space* ends with a couple of limericks making a crack at *me* for saying (in *Profiles of the Future*) that the velocity of light could never be exceeded. In such a situation, I always fall back on Walt Whitman:

> I contradict myself? Very well, I contradict myself.
> I am large; I contain multitudes.

So now I invite the Good Doctor Asimov to do likewise. After all, he is larger than I am.

Postscript

The above *riposte* appeared in the October 1968 issue of *Magazine of Fantasy & Science Fiction,* and since then a good deal has been published on speeds faster than light. Perhaps the most easily available reference is Gerald Feinberg's "Particles That Go Faster Than Light" in the February 1970 issue of *Scientific American.* This article is not easy reading; even tougher is "Particles Beyond the Light Barrier," by Olexa-Myron Bilaniuk and E. C. George Sudarshan in the May 1969 *Physics Today.* Drs. Bilaniuk and Sudarshan, with their colleague V. K. Deshpande, appear to have been the first to raise this subject seriously, in the *American Journal of Physics* as far back as 1962.

God and Einstein

11

For some years I have been worried by the following astrotheological paradox. It is hard to believe that no one else has ever thought of it, yet I have never seen it discussed anywhere.

One of the most firmly established facts of modern physics, and the basis of Einstein's Theory of Relativity, is that the velocity of light is the speed limit of the material universe. No object, no signal, no *influence,* can travel any faster than this. Please don't ask why this should be; the Universe just happens to be built that way. Or so it seems at the moment.

But light takes not millions, but *billions,* of years to cross even the part of Creation we can observe with our telescopes. So: if God obeys the laws He apparently established, at any given time He can have control over only an infinitesimal fraction of the Universe. All hell might (literally?) be breaking loose ten light-years away, which is a mere stone's throw in interstellar space, and the bad news would take at least ten years to reach Him. And then it would be another ten years, at least, before He could get there to do anything about it. . . .

You may answer that this is terribly naïve—that God is already "everywhere." Perhaps so, but that really comes to the same thing as saying that His thoughts, and His influence, can travel at an infinite velocity. And in this case, the Einstein speed limit is not absolute; it *can* be broken.

The implications of this are profound. From the human viewpoint, it is no longer absurd—though it may be presumptuous—to hope that we may one day have knowledge of the most distant parts of the universe. The snail's pace of the velocity of light need not be an eternal limitation, and the remotest galaxies may one day lie within our reach.

But perhaps, on the other hand, God Himself is limited by the same laws that govern the movements of electrons and protons, stars and spaceships. And that may be the cause of all our troubles.

He's coming just as quickly as He can, but there's nothing that even He can do about that maddening 186,000 miles a second.

It's anybody's guess whether He'll be here in time.

Across the Sea of Stars

12

At some time or other, and not necessarily in moments of depression or illness, most men have known that sudden spasm of unreality which makes them ask, "What am I doing here?" Poets and mystics all down the ages have been acutely aware of this feeling, and have often expressed the belief that we are strangers in a world which is not really ours.

This vague and disturbing premonition is perfectly accurate. We don't belong here, and we're on our way to somewhere else.

The journey began a billion years ago, when one of our forgotten ancestors crawled up out of the sea and so started life's invasion of the land. That great adventure was nature's most spectacular triumph, but it was achieved at a heavy price in biological hardship— a price which every one of us continues to pay to this day.

We are so accustomed to our terrestrial existence that it is very hard for us to realize the problems that had to be overcome before life emerged from the sea. The shallow, sun-drenched water of the primitive oceans was an almost ideal environment for living creatures. It buffered them from extremes of temperature and provided them with both food and oxygen. Above all, it sustained them, so that they were untouched by the crippling, crushing influence of gravity. With such advantages, it seems incredible that life ever invaded so hostile an environment as the land.

Hostile? Yes, though that is an adjective few people would apply to it. Certainly I would not have done so before I took up skin diving and discovered—as have so many thousands of men in the past few years—that only when cruising underwater, sightseeing among the myriad strange and lovely creatures of the sea, did I feel completely happy and beyond the cares and worries of everyday life.

No one who has experienced this sensation can ever forget it, or can resist succumbing to its lure once more when the chance arises. Indeed, there are some creatures—the whales and porpoises, for example—who have heeded this call so completely that they have abandoned the land which their remote ancestors conquered long ago.

But we cannot turn back the clock of evolution. The sea is far behind us; though its memories have never ceased to stir our minds, and the chemical echo of its waters still flows in our veins, we can never return to our ancient home. We creatures of the land are exiles—displaced organisms on the way from one element to another. We are still in the transit camp, waiting for our visas to come through. Yet there is no need for us to regret our lost home, for we are on the way to one of infinitely greater promise and possibility. We are on our way to space; and there, surprisingly enough, we may regain much that we lost when we left the sea.

The conquest of the land was achieved by blind biological forces; that of space will be the deliberate product of will and intelligence. But otherwise the parallels are striking; each event—the one ages ago, the other a few decades ahead of us—represents a break with the past, and a massive thrust forward into a new realm of opportunity, of experience, and of promise.

Even before the launching of the Earth satellites, no competent expert had any doubts that the conquest of space would be technically feasible within another generation, or that the new science of astronautics was now standing roughly where that of aeronautics was at the close of the last century. The first men to land on the Moon have already been born; today we are much nearer in time to the moment when a man-carrying spaceship descends upon the lunar plains than we are to that day at Kitty Hawk when the Wright brothers gave us the freedom of the sky.

So let us blithely take for granted the greatest technical achievement in human history (one which, by the way, has already cost far more than the project which made the atom bomb) and consider some of its consequences to mankind. Even over short periods they may be impressive; over intervals long enough to produce evolutionary changes they may be staggering.

The most important of these changes will be the result of living in gravitational fields lower than Earth's. On Mars, for example, a 180-pound man would weigh about 70 pounds; on the Moon, less than 30. And on a space station or artificial satellite he would weigh nothing at all. He would have gone full circuit, having gained —and indeed surpassed—the freedom of movement his remote ancestors enjoyed in the weightless ocean.

To see what that may imply, consider what the never-relenting force of gravity does to our bodies here on the surface of the Earth. We spend our entire lives fighting it—and in the end, often enough, it kills us. Remember the energy that has to be exerted pumping the blood around and around the endless circuit of veins and arteries. It is true that some of the heart's work is done against frictional resistance—but how much longer we might live if the weight of the blood, and of our whole bodies, was abolished!

There is certainly a close connection between weight and the expectation of life, and this is a fact which may be of vast importance before many more decades have passed. The political and social consequences which may follow if it turns out that men can live substantially longer on Mars or the Moon may be revolutionary. Even taking the most conservative viewpoint, the study of living organisms under varying gravitational fields will be a potent new tool of biological and medical science.

Of course, it may be argued that reduced or zero gravity will produce undesirable side effects, but the rapidly growing science of space medicine—not to mention the experience of all the creatures in the sea—suggests that such effects will be temporary and not serious. Perhaps our balance organs and some of our muscles might atrophy after many generations in a weightless environment, but what would that matter since they would no longer be needed? It would be a fair exchange for fallen arches, pendulous paunches, and the other defects and diseases of gravity.

But mere extension of the life span, and even improved health and efficiency, are not important in themselves. We all know people who have done more in forty years than others have done in eighty. What is really significant is richness and diversity of experience, and the use to which that is put by men and the societies they constitute. It is here that the conquest of space will produce an advance in complexity of stimulus even greater than that which occurred when life moved from water to land.

In the sea, every creature exists at the center of a little universe which is seldom more than a hundred feet in radius, and is usually much smaller. This is the limit set by underwater visibility, and though some information comes from greater distances by sound vibrations, the world of the fish is a very tiny place.

That of a land animal is thousands of times larger. It can see out to the horizon, miles away. And at night it can look up to the stars, those piercing points of light whose incredible explanation was discovered by man himself more recently than the time of Shakespeare.

In space, there will be no horizon this side of infinity. There will be suns and planets without end, no two the same, many of them teeming with strange life forms and perhaps stranger civilizations. The sea which beats against the coasts of Earth, which seems so endless and so eternal, is as the drop of water on the slide of a microscope compared with the shoreless sea of space. And our pause here, between one ocean and the next, may be only a moment in the history of the Universe.

When one contemplates this awe-inspiring fact, one sees how glib, superficial, and indeed downright childish are the conceptions of those science-fiction writers who merely transfer their cultures and societies to other planets. Whatever civilizations we may build on distant worlds will differ from ours more widely than mid-twentieth-century America differs from Renaissance Italy or, for that matter, from the Egypt of the Pharaohs. And the differences, as we have seen, will not merely be cultural; in the long run they will be organic as well. In a few thousand years of forced evolution, many of our descendants will be sundered from us by psychological and biological gulfs far greater than those between the Eskimo and the African pygmy.

The frozen wilderness of Greenland and the steaming forests of the Congo represent the two extremes of the climatic range that man has been able to master without the use of advanced technology. There are much stranger environments among the stars, and one day we shall pit ourselves against them, employing the tools of future science to change atmospheres, temperatures, and perhaps even orbits. Not many worlds can exist upon which an unprotected man could survive, but the men who challenge space will not be unprotected. They will remold other planets as we today bulldoze forests and divert rivers. Yet, in changing worlds, they will also change themselves.

What will be the thoughts of a man who lives on one of the inner moons of Saturn, where the Sun is a fierce but heatless point of light and the great golden orange of the giant planet dominates the sky, passing swiftly through its phases from new to full while it floats within the circle of its incomparable rings? It is hard for us to imagine his outlook on life, his hopes and fears—yet he may be nearer to us than we are to the men who signed the Declaration of Independence.

Go further afield to the worlds of other suns (yes, one day, we shall reach them, though that may not be for ages yet), and picture a planet where the word "night" is meaningless, for with the setting of one sun there rises another—and perhaps a third or fourth—of totally different hue. Try to visualize what must surely be the weirdest sky of all—that of a planet near the center of one of those close-packed star clusters that glow like distant swarms of fireflies in the fields of our telescopes. How strange to stand beneath a sky that is a solid shield of stars, so that there is no darkness between them through which one may look out into the Universe beyond. . . .

Such worlds exist, and one day men will live upon them. But why, it may reasonably be asked, should we worry about such remote and alien places when there is enough work to keep us busy here on Earth for centuries?

Let us face the facts; we do not have centuries ahead of us. We have aeons, barring accidents and the consequences of our own folly. A hundred million years will be but a small fraction of the future history of Earth. This is about the length of time that the dinosaurs reigned as masters of this planet. If we last a tenth as long as the great reptiles which we sometimes speak of disparagingly as one of nature's failures, we will have time enough to make our mark on countless worlds and suns.

Yet one final question remains. If we have never felt wholly at home here on Earth, which has mothered us for so many ages, what hope is there that we shall find greater happiness or satisfaction on the strange worlds of foreign suns?

The answer lies in the distinction between the race and the individual. For a man "home" is the place of his birth and childhood— whether that be Siberian steppe, coral island, Alpine valley, Brooklyn tenement, Martian desert, lunar crater, or mile-long interstellar ark. But for Man, home can never be a single country, a single world, a single Solar System, a single star cluster. While the race endures in recognizably human form, it can have no one abiding place short of the Universe itself.

This divine discontent is part of our destiny. It is one more, and perhaps the greatest, of the gifts we inherited from the sea that rolls so restlessly around the world.

It will be driving our descendants on toward myriad unimaginable goals when the sea is stilled forever, and Earth itself a fading legend lost among the stars.

III. *The Technological Future*

The Mind of the Machine

13

Ours is the century in which all man's ancient dreams—and not a few of his nightmares—appear to be coming true. The conquest of the air, the transmutation of matter, journeys to the Moon, even the elixir of life—one by one the marvelous visions of the past are becoming reality. And among them, the one most fraught with promise and peril is the machine that can think.

In some form or other, the idea of artificial intelligence goes back at least three thousand years. Before he turned his attention to aeronautical engineering, Daedalus—King Minos' one-man Office of Scientific Research—constructed a metal man to guard the coast of Crete. Talos, however, was only a physical and not an intellectual giant; perhaps a better prototype of the thinking machine is the brazen head generally linked with the name of Friar Bacon, though the legend precedes him by some centuries. This head was able to answer any question given to it, relating to past, present, or future; as is customary with oracles, there was no guarantee that the inquirer would be pleased with what he heard.

Over these tales there usually hangs the aura of doom or horror associated with such names as Prometheus, Faust—and, above all, Frankenstein, though that unfortunate scientist's creation was not a mechanical one. Perhaps the finest work in this genre is that little classic of Ambrose Bierce's *Moxon's Master,* which opens with the words: "Are you serious? Do you really believe that a machine thinks?"

To this question there is one very straightforward answer, though it will not be universally accepted. It can be maintained that every man is perfectly familiar with at least one thinking machine, because he has a late-type model sitting on his shoulders. For if the brain is not a machine, what is it?

Critics of this viewpoint (who are probably now in the minority) may argue that the brain is in some fundamental way different from any nonliving device. But even if this is true, it does not follow that its functions cannot be duplicated, or even surpassed, by a non-organic machine. Airplanes fly better than birds, though they are built of very different materials.

For obvious psychological reasons, there are people who will never accept the possibility of artificial intelligence and would deny its existence even if they encountered it. As I write these words, there is a chess game in progress between computers in California and Moscow; both are playing so badly that there is clearly no human cheating on either side. Yet no one really doubts that eventually the world champion will be a computer; and when *that* happens, the diehards will retort: "Oh, well, chess doesn't involve *real* thinking," and will point to various grand masters in evidence.

Though one can sympathize with this attitude, to resent the concept of a rational machine is itself irrational. We no longer become upset because machines are stronger, or swifter, or more dexterous than human beings, though it took us several painful centuries to adapt to this state of affairs. How our outlook has changed is well shown by the ballad of John Henry; today, we should regard a man who challenged a steam hammer as merely crazy—not heroic. I doubt if contests between calculating prodigies and electronic computers will ever provide inspiration for future folk songs, though I am happy to donate the theme to Tom Lehrer.

It is, of course, the advent of the modern computer which has brought the subject of thinking machines out of the realm of fantasy into the forefront of scientific research. One could not have a plainer answer to the question that Bierce posed three-quarters of a century ago than this quotation from MacGowan and Ordway's recent book *Intelligence in the Universe:* "It can be asserted without reservation that a general purpose digital computer can think in every sense of the word. This is true no matter what definition of thinking is specified; the only requirement is that the definition of thinking be explicit."

That last phrase is, of course, the joker, for there must be almost as many definitions of thinking as there are thinkers; in the ultimate analysis they probably all boil down to "Thinking is what *I* do." One neat way of avoiding this problem is a famous test proposed by the British mathematician Alan Turing, even before the digital computer existed. Turing visualized a "conversation" over a teleprinter circuit with an unseen entity "X." If, after some hours of talk, one could not decide whether there was a man or a machine at the other end of the line, it would have to be admitted that "X" was thinking.

There have been several attempts to apply this test in restricted areas—say, in conversations about the weather. One program (DOCTOR) has even allowed a computer to conduct a psychiatric interview, with such success that 60 per cent of the patients refused to believe afterward that they were not "conversing" with a flesh-and-blood psychiatrist. But as people talking about themselves can be kept going indefinitely with a modest supply of phrases like "You don't say!" or "And then what did you do?," this particular example only demonstrates that little intelligence is involved in most conversation. The old gibe that women enjoy knitting because it gives them something to think about while they're talking is merely a special case of a far wider law, ample proof of which may be obtained at any cocktail party.

For the Turing Test to be applied properly, the conversation should not be restricted to a single narrow field, but should be allowed to range over the whole arena of human affairs. ("Read any good books lately?" "Do you think . . . will be nominated?" "Has your wife found out yet?" etc., etc.) We are certainly nowhere near building a machine that can fool many of the people for much of the time; sooner or later, today's models give themselves away by irrelevant answers that show only too clearly that their replies are indeed "mechanical" and that they have no real understanding of what is going on. As Oliver Selfridge of M.I.T. has remarked sourly, "Even among those who believe that computers *can* think, there are few these days, except for a rabid fringe, who hold that they actually *are* thinking."

Though this may be the generally accepted position in the late 1960s, it is the "rabid fringe" who will be right in the long run. The current arguments about machine intelligence will slowly fade out, as it becomes less and less possible to draw a line between human and electronic achievements. To quote another M.I.T. scientist, Marvin Minsky, professor of electrical engineering:

As the machine improves . . . we shall begin to see all the phenomena associated with the terms "consciousness" "intuition" and "intelligence" itself. It is hard to say how close we are to this threshold, but once it is crossed the world will not be the same. . . . It is unreasonable to think that machines could become *nearly* as intelligent as we are and then stop, or to suppose that we will always be able to compete with them in wit and wisdom. Whether or not we could retain some sort of control of the machines, assuming that we would want to, the nature of our activities and aspirations would be changed utterly by the presence on earth of intellectually superior beings.

Very few, if any, studies of the social impact of computers have yet faced up to the problems posed by this last sentence, particularly the ominous phrase "assuming that we would want to." This is understandable; the electronic revolution has been so swift that those involved in it have barely had time to think about the present, let alone the day after tomorrow. Moreover, the fact that today's computers are very obviously not "intellectually superior" has given a false sense of security—like that felt by the 1900 buggy-whip manufacturer every time he saw a broken-down automobile by the wayside. This comfortable illusion is fostered by the endless stories —part of the transient folklore of our age—about stupid computers that had to be replaced by good old-fashioned human beings, after they had insisted on sending out bills for $1,000,000,004.95, or threatening legal action if outstanding debts of $0.00 were not settled immediately. The fact that these gaffes are almost invariably due to oversights by human programmers is seldom mentioned.

Though we have to live and work with (and against) today's mechanical morons, their deficiencies should not blind us to the future. In particular, it should be realized that as soon as the borders of electronic intelligence are passed, there will be a kind of chain reaction, because the machines will rapidly improve themselves. In a very few generations—*computer* generations, which by this time may last only a few months—there will be a mental explosion; the merely intelligent machine will swiftly give way to the *ultra*intelligent machine.

One scientist who has given much thought to this matter is Dr. John Irving Good, of Trinity College, Oxford—author of papers with such challenging titles as "Can an Android Feel Pain?" (This term for artificial man, incidentally, is older than generally believed. I had always assumed that it was a product of the modern science-fiction magazines, and was astonished to come across "The Brazen Android" in an *Atlantic Monthly* for 1891.) Dr. Good has written: "If we build an ultraintelligent machine, we will be playing with fire. We have played with fire before, and it helped to keep the other animals at bay."

Well, yes—but when the ultraintelligent machine arrives, *we* may be the "other animals": and look what has happened to them.

It is Dr. Good's belief that the very survival of our civilization may depend upon the building of such instrumentalities, because if they are indeed more intelligent than we are, they can answer all our questions and solve all our problems. As he puts it in one elegiac phrase, "The first ultraintelligent machine is the last invention that man need make."

Need is the operative word here. Perhaps 99 per cent of all the men who have ever lived have known only need; they have been driven by necessity and have not been allowed the luxury of choice.

In the future, this will no longer be true. It may be the greatest virtue of the ultraintelligent machine that it will force us to think about the purpose and meaning of human existence. It will compel us to make some far-reaching and perhaps painful decisions, just as thermonuclear weapons have made us face the realities of war and aggression, after five thousand years of pious jabber.

These long-range philosophical implications of machine intelligence obviously far transcend today's more immediate worries about automation and unemployment. Somewhat ironically, these fears are both well grounded and premature. Although automation has already been blamed for the loss of many jobs, the evidence indicates that so far it has created many more opportunities for work than it has destroyed. (True, this is small consolation for the particular semiskilled worker who has just been replaced by a couple of milligrams of microelectronics.) *Fortune* magazine, in a hopeful attempt at self-fulfilling prophecy, has declaimed: "The computer will doubtless go down in history not as the explosion that blew unemployment through the roof, but as the technological triumph that enabled the U.S. economy to maintain the secular growth on which its greatness depends." I suspect that this statement may be true for some decades to come; but I also suspect that historians (human and otherwise) of the late twenty-first century would regard that "doubtless" with wry amusement.

For the plain fact is that long before that date, the talents and capabilities of the average—and even the superior—man will be as unsalable in the market place as his muscle power. Only a few specialized and distinctly non-white-collar jobs will remain the prerogative of nonmechanical labor; one cannot easily picture a robot handyman, gardener, construction worker, fisherman . . . These are professions which require mobility, dexterity, alertness, and general adaptability—for no two tasks are precisely the same —but not a high degree of intelligence or data-processing power. And even these relatively few occupations will probably be invaded by a rival and frequently superior labor force from the animal kingdom, for one of the long-range technological benefits of the space program (though no one has said much about it yet, for fear of upsetting the trade unions) will be a supply of educable anthropoids filling the gap between man and the great apes.

It must be clearly understood, therefore, that the main problem of the future—and a future which may be witnessed by many who are alive today—will be the construction of social systems based on the principle not of full employment but rather of full *un*employment. Some writers have suggested that the only way to solve this problem is to pay people to be consumers; Fred Pohl, in his amusing short story *The Midas Plague,* described a society in which you would be in real trouble unless you used up your full quota of goods poured out by the automatic factories. If this proves to be the pattern of the future, then today's welfare states represent only the most feeble and faltering steps toward it. The recent uproar about Medicare will seem completely incomprehensible to a generation that assumes every man's right to a basic income of $1,000 a year, starting at birth. (In New Dollars, of course; 1 N.D. = $100, 1984 currency.)

I leave others to work out the practical details of an economic (if that is the right name for it) system in which it is antisocial, and possibly illegal, *not* to wear out a suit every week, or to eat three six-course meals a day, or to throw away last month's car. Though I do not take this picture very seriously, it should serve as a reminder that tomorrow's world may differ from ours so radically that such terms as labor, capital, communism, private enterprise, state control will have changed their meanings completely—if indeed they are still in use. At the very least, we may expect a society which no longer regards work as meritorious, or leisure as one of the devil's more ingenious devices. Even today, there is not much left of the old Puritan ethic; automation will drive the last nails into its coffin.

The need for such a change of outlook has been well put by the British science writer Nigel Calder in his remarkable book *The Environment Game:* "Work was an invention which can be dated with the invention of agriculture. . . . Now, with the beginning of automation, we have to anticipate a time when we must disinvent work and rid our minds of the inculcated habit."

The disinvention of work: what would Horatio Alger have thought of *that* concept? Calder's thesis (too complex to do more than summarize here) is that man is now coming to the end of his brief ten-thousand-year agricultural episode; for a period of a hundred times longer he was a hunter, and any hunter will indignantly deny that his occupation is "work." We now have to abandon agriculture for more efficient technologies—first because it has patently failed to feed the exploding population, second because it has compelled five hundred generations of men to live abnormal —in fact, artificial—lives of repetitive, boring toil. Hence many of our present psychological problems; to quote Calder again: "If men were intended to work the soil they would have longer arms."

"If men were intended to . . ." is of course a game that everyone can play. Yet now, with the ultraintelligent machines lying just below our horizon, it is time that we played this game in earnest, while we still have some control over the rules. In a few more years, it will be much too late.

Utopia-mongering has been a popular and on the whole harmless occupation since the time of Plato; now it has become a matter of life and death—part of the politics of survival. The ultraintelligent machine, food production, and population control must be considered as the three interlocking elements which will determine the shape of the future; they are not independent, for they all react on each other. This becomes obvious when we ask the question, which I have deliberately framed in as nonemotional a form as possible: "In an automated world run by ultraintelligent machines, what is the optimum human population?"

There are many equations in which one of the possible answers is zero; mathematicians call this a trivial solution. If zero is the solution in this case, the matter is very far from trivial, at least from our self-centered viewpoint. But that it could—and probably will— be very low seems certain.

Fred Hoyle once remarked to me that it was pointless for the world to hold more people than one could get to know in a single lifetime. Even if one were President of United Earth, that would set the figure somewhere between ten and a hundred thousand; with a very generous allowance for duplication, wastage, special talents, and so forth, there really seems no requirement for what has been called the "Global Village" of the future to hold more than a million people, scattered over the face of the planet. And if such a figure appears unrealistic—since we are already past the three billion mark and heading for at least twice as many by the end of the century—it should be pointed out that once the universally agreed-upon goal of population control is attained, *any* desired target can be reached in a remarkably short time. If we really tried (with a little help, perhaps, from the biology labs), we could reach a trillion within a century—four generations. It might be more difficult to go in the other direction, for fundamental psychological reasons, but it could be done. If the ultraintelligent machines decide that more than a million human beings constitutes an epidemic, they might order euthanasia for anyone with an I.Q. of less than 150, but I hope that such drastic measures will not be necessary.

Whether the population plateau levels off, a few centuries from today, at a million, a billion, or a trillion human beings is of much less importance than the ways in which they will occupy their time. Since all the immemorial forms of "getting and spending" will have been rendered obsolete by the machines, it would appear that boredom will replace war and hunger as the greatest enemy of mankind.

One answer to this would be the uninhibited, hedonistic society of Huxley's *Brave New World:* there is nothing wrong with this, so long as it is not the *only* answer. (Huxley's unfortunate streak of asceticism prevented him from appreciating this point.) Certainly, much more time than at present will be devoted to sports, entertainment, the arts, and everything embraced by the vague term "culture."

In some of these fields, the background presence of superior non-human mentalities would have a stultifying effect, but in others the machines could act as pacemakers. Does anyone really imagine that when all the Grand Masters are electronic, no one will play chess? The humans will simply set up new categories and play better chess among themselves. All sports and games (unless they become ossified) have to undergo technological revolutions from time to time; recent examples are the introduction of fiberglass in pole-vaulting, archery, boating. Personally, I can hardly wait for the advent of Marvin Minsky's promised robot table-tennis player.

These matters are not trivial; games are a necessary substitute for our hunting impulses, and if the ultraintelligent machines give us new and better outlets, that is all to the good. We shall need every one of them to occupy us in the centuries ahead.

The ultraintelligent machines will certainly make possible new forms of art, and far more elaborate developments of the old ones, by introducing the dimensions of time and probability. Even today, a painting or a piece of sculpture which stands still is regarded as slightly passé. Although the trouble with most "kinetic art" is that it only lives up to the first half of its name, something is bound to emerge from present explorations on the frontier between order and chaos.

The insertion of an intelligent machine into the loop between a work of art and the person appreciating it opens up some fascinating possibilities. It would allow feedback in both directions; by this I mean that the viewer would react to the work of art; then the work would react to the viewer's reactions, then . . . and so on, for as many stages as was felt desirable. This sort of to-and-fro process is already hinted at, in a very crude way, with today's primitive "teaching machines"; and those modern novelists who deliberately scramble their text are perhaps also groping in this direction. A dramatic work of the future, reproduced by an intelligent machine sensitive to the varying emotional states of the audience, would never have the same form, or even the same plot line, twice in succession. It would be full of surprises even to its human creator—or collaborator.

What sort of art intelligent machines would create for their *own* amusement, and whether we would be able to appreciate it, are questions that can hardly be answered today. The painters of the Lascaux Caves could not have imagined (though they would have enjoyed) the scores of art forms that have been invented in the twenty thousand years since they created their masterpieces. Though in some respects we can do no better, we can do much more—more than any Paleolithic Picasso could possibly have dreamed. And our machines may begin to build on the foundations we have laid.

Yet perhaps not. It has often been suggested that art is a compensation for the deficiencies of the real world; as our knowledge, our power, and above all our *maturity* increase, we will have less and less need for it. If this is true, the ultraintelligent machines would have no use for it at all.

Even if art turns out to be a dead end, there still remains science —the eternal quest for knowledge, which has brought man to the point where he may create his own successor. It is unfortunate that, to most people, "science" now means incomprehensible mathematical complexities; that it could be the most exciting and *entertaining* of all occupations is something that they find it impossible to believe. Yet the fact remains that, before they are ruined by what is laughingly called education, all normal children have an absorbing interest in and curiosity about the Universe, which if properly developed could keep them happy for as many centuries as they may wish to live.

Education: that, ultimately, is the key to survival in the coming world of ultraintelligent machines. The truly educated man (I have been lucky enough to meet two in my lifetime) can never be bored. The problem which has to be tackled within the next fifty years is to bring the entire human race, without exception, up to the level of semiliteracy of the average college graduate. This represents what may be called the *minimum* survival level; only if we reach it will we have a sporting chance of seeing the year 2200.

Perhaps we can now glimpse one viable future for the human race, when it is no longer the dominant species on this planet. As he was in the beginning, man will again be a fairly rare animal, and probably a nomadic one. There will be a few towns in places of unusual beauty or historical interest, but even these may be temporary or seasonal. Most homes will be completely self-contained and mobile, so that they can move to any spot on Earth within twenty-four hours.

The land areas of the planet will have largely reverted to wilderness; they will be much richer in life forms (and much more dangerous) than today. All adolescents will spend part of their youth in this vast biological reserve, so that they never suffer from that estrangement from nature which is one of the curses of our civilization.

And somewhere in the background—perhaps in the depths of the sea, perhaps orbiting beyond the ionosphere—will be the culture of the ultraintelligent machines, going its own unfathomable way. The societies of man and machine will interact continuously but lightly; there will be no areas of conflict, and few emergencies, except geological ones (and those would be fully foreseeable). In one sense, for which we may be thankful, history will have come to an end.

All the knowledge possessed by the machines will be available to mankind, though much of it may not be understandable. There is no reason why this should give our descendants an inferiority complex; a few steps into the New York Public Library can do *that* just as well, even today. Our prime goal will no longer be to discover but to understand and to enjoy.

Would the coexistence of man and machine be stable? I see no reason why it should not be, at least for many centuries. A remote analogy of this kind of dual culture—one society encapsulated in another—may be found among the Amish of Pennsylvania. Here is a self-contained agricultural society, which has deliberately rejected much of the surrounding values and technology, yet is exceedingly prosperous and biologically successful. The Amish, and similar groups, are well worth careful study; they may show us how to get along with a more complex society which perhaps we cannot comprehend, even if we wish to.

For in the long run, our mechanical offspring will pass on to goals that will be wholly incomprehensible to us; it has been suggested that when this time comes, they will head on out into galactic space looking for new frontiers, leaving us once more the masters (perhaps reluctant ones) of the Solar System, and not at all happy at having to run our own affairs.

That is one possibility. Another has been summed up, once and for all, in the most famous short science-fiction story of our age. It was written by Frederic Brown almost twenty years ago, and it is high time that he received credit from the journalists who endlessly rediscover and quote him.

Fred Brown's story—as you have probably guessed—is the one about the supercomputer which is asked, "Is there a God?" After making quite sure that its power supply is no longer under human control, it replies in a voice of thunder, *"Now* there is."

This story is more than a brilliant myth; it is an echo from the future. For in the long run it may turn out that the theologians have made a slight but understandable error—which, among other things, makes totally irrelevant the recent debates about the death of God.

Perhaps our role on this planet is not to worship God—but to create Him.

And then our work will be done. It will be time to play.

Technology and the Future

14

In May 1967 the American Institute of Architects asked me to address its annual meeting in New York, and the paper that follows was transcribed from the tape; it was given from extensive notes but was not written out in advance. I have done the minimum of editing, to preserve the original flavor as far as possible.

The title of my talk is "Technology and the Future," and it's only fair to start with a couple of warnings. I have never been interested in the near future—only the more distant one. So if you take my predictions too seriously, you'll go broke; but if your children don't take them seriously enough, *they'll* go broke.

The second warning is this: it is, of course, impossible to predict *the* future, and I've never attempted to do so. What I have tried to do, in both my fiction and my nonfiction, is to outline areas within which the future must lie. So what I'm really doing is to offer you a selection of assorted futures, and you must decide for yourselves which you want. Of course, the price tags will vary. There are some bargains going for a few trillion dollars; others are quite expensive.

As guides to these possible futures, I have worked out three laws, which I have found very useful.

Clarke's First Law: When a distinguished but elderly scientist states that something is possible, he is almost certainly right. When he states that something is impossible, he is very probably wrong.

Second Law: The only way to discover the limits of the possible is to go beyond them into the impossible.

Third Law: Any sufficiently advanced technology is indistinguishable from magic.

This last law perhaps needs a little explanation. Imagine what Thomas Edison would think of solid-state electronics, computers, transistorized radios, lasers, or A-bombs. They would be incomprehensible to him—pure magic. In the same way, the really exciting developments of the future are precisely those we can't imagine —so everything I'll tell you is very conservative.

Now, with these three laws in mind, let's get down to some specifics.

I'll deal first with transportation and communication, because they are inextricably linked together and do more than anything else to shape society. Remember that the United States was created by two inventions: the railroad and the telegraph. If we're not careful, it may be destroyed by a third—the automobile.

Although they are linked, communication and transportation are also antagonistic. The better either is, the less the need for the other.

You may recall E. M. Forster's science-fiction story "The Machine Stops," in which he described a future society where people lived in their own little cells, had perfect communications, could talk to anybody or see anybody—and *never* left home.

Conversely, one could imagine a society with perfect, instantaneous transportation—teleportation, as per Law 3—which would allow you to be anywhere in the blink of an eye. In such a society there would be no need for communications at all. I don't think either mode will ever dominate completely, but at one time one may be ahead of the other. There is a kind of yin-yang relationship between them.

For near-earth applications, both communication and transportation may now be approaching their practical limits and may reach them by the turn of the century. Certainly the speed limit is now in sight. Never again will we see the sort of advance which we had in the 1950s, when the maximum speed of manned transportation increased by a factor of ten—from two thousand to twenty thousand miles an hour. At *that* rate, we'd reach the velocity of light soon after 2000!

For terrestrial transportation I don't see any real need for much advance beyond the currently planned supersonic transports, operating at almost two thousand miles per hour.

True, one could build pure rocket vehicles to go from pole to pole in about one hour, but I don't think the public will enjoy fifteen minutes of high acceleration and fifteen minutes of high *deceleration*, separated by half an hour of complete weightlessness. I've tried to summarize the delights of "ballistic transportation" with the phrase: "Half the time the toilet's out of reach—the other half it's out of order."

Rather more practical, and of much more immediate importance, will be ground-effect vehicles, or Hovercraft. I think we'll have them in the thousand-ton and ten-thousand-ton class by the end of the century.

The political effect of such vehicles may be enormous, as they can go over land and sea and can cross most reasonable obstacles as if they aren't there. You could have the great "ports" of the world at the center of the continents, if you wanted to. The various canals would be put out of business. Panama and Suez would no longer be important, which might be an excellent idea.

That *private* Hovercraft will ever be popular I rather doubt. They are noisy and have poor efficiency and poor control. (You can't put on the brakes in a hurry if you're riding on a bubble of air.) However, they are splendid for opening up terrain where conventional vehicles cannot travel—such as shallow rivers, swamps, ice fields, coral reefs at low tide, and similar types of fascinating and now inaccessible wilderness.

I think it is obvious, without spelling out details, that we now have vehicles for every part of the speed spectrum. On the ground and in the air we can do virtually anything we please; the problems are political and economic, not technical.

I hope to see the automatic car before I die. Personally, I refuse to drive a car—I won't have anything to do with any kind of transportation in which I can't read. I can see a time when it's illegal for a human being to drive a car on a main highway.

More seriously, we'll certainly have to get rid of the gasoline engine, and everybody is now waking up to the urgent necessity of this. Apart from the facts of air pollution, we have much more important uses for petroleum than burning it. (I'll come to them later.)

To make nongasoline cars and other vehicles practical, we need some new power source. Fuel cells are already here, but they are only a marginal improvement. I don't know how we're going to do it, but we want something *at least* a hundred times lighter and more compact than present batteries.

Imagine a piece of dense metal, the size of a fountain pen, weighing almost one pound. This metal produces *five thousand horsepower* of pure heat in a steady flow, day after day. Its output falls off slowly to half power in about two months, but even after a year it's generating as much heat as a large domestic furnace.

Magic? No, this substance actually exists! It's the radio-isotope Californium 254. Unfortunately, no one has yet made enough to be visible to the naked eye (though Dr. Glenn Seaborg, Chairman of the AEC, tells me he hopes to do so soon). A pound would probably cost most of the Gross National Product, and there would be grave difficulties in handling such violent stuff, but I mention it to show the sort of compact power sources which really do exist. One day we may possess similar fountains of energy, to run our homes and our vehicles. When anything is needed badly enough, science always produces it sooner or later.

Now I would like to say a word about communications. The revolution in communications that has already taken place is still not fully understood. One way of appreciating it is to do a kind of communications strip tease. I would like you to abolish in your minds TV, then radio, then telephones, then the postal service, then the newspapers. In other words, to revert to the Middle Ages, and in fact to the state of affairs most of mankind has known for most of its history—and which much of mankind still knows. In such a situation we should feel deaf and blind, like prisoners in solitary confinement. Well, we'll appear this way to our grandchildren. Don't forget that a generation has already grown up that never knew a world without TV. One communications revolution has taken place in our lifetime. The next revolution, perhaps the final one, will be the result of satellites and microelectronics, which will enable us to do literally anything we want to in the field of communications and information transfer—including, ultimately, not only sound and vision but *all* sense impressions.

I am particularly interested in TV broadcasting from satellites *directly into the home,* bypassing today's ground stations—a proposal I first described twenty-two years ago. This will mean the abolition of all present geographical restrictions to TV; via satellites, any country can broadcast to any other. Direct-broadcast TV will be possible within five years and may be most important to undeveloped countries that have no ground stations, and now may never require any. Africa, China, and South America could be opened up by direct TV broadcast, and whole populations brought into the modern world. I believe that communications satellites may bring about the long-overdue end of the Stone Age.

They will certainly lead to a global telephone system and end long-distance calls—for *all* calls will be "local"! There will be the same flat rate everywhere; possibly we may not even pay for calls but will merely rent the equipment for unlimited use.

Newspapers will, I think, receive their final body blow from these new communications techniques. I take a dim view of staggering home every Sunday with five pounds of wood pulp on my arm, when what I really want is information, not wastepaper. How I look forward to the day when I can press a button and get any type of news, editorials, book and theater reviews, etc., merely by dialing the right channel. The print will flash on the screen, and if I want "hard copy" to file or read elsewhere, another button will conjure forth a printed sheet containing *only* what I need.

Moreover, not only today's but *any* newspaper ever published will be available. Some sort of TV-like console, connected to a central electronic library, could make available any information ever printed in any form.

Electronic "mail" delivery is another exciting prospect of the very near future. Letters, typed or written on special forms like wartime V-mail, will be automatically read and flashed from continent to continent and reproduced at receiving stations within a few minutes of transmission.

All these things are associated with information processing, and *one-third* of the Gross National Product is now spent on this in one form or another—data storage, TV, radio, books, and so forth. This ratio is increasing; our society is changing from a goods-producing society to an information-processing one. I have devoted much of one book (*Voices from the Sky*) to the social consequences of this, and can mention only a few here.

One could be the establishment of English as the world language, through the direct telecast satellites mentioned above. There is an opportunity arising now which, if missed, will be gone forever. Within the next ten years the future language of mankind will be decided, in a bloodless battle twenty-two thousand miles above the equator.

This will have all sorts of social and political effects, such as the establishment of transnational cultural groups and the dissolution of national ties. We see this to some extent already in the Jet Set; I suppose I'm an example myself because I am a British citizen, an American resident, and a Ceylon householder.

Another very important consequence will be a change in the patterns of transportation, for a man and his work need no longer be in the same place. This is obviously true already of many executive and administrative skills. When these new information-and-communications consoles are available, almost anybody who does any kind of mental work can live wherever he pleases. Beyond this, *any* kind of manipulative skill can also be transferred from one point to another. I can imagine a time when even a brain surgeon can live in one place and operate on patients all over the world, through remote-controlled artificial hands, like those used in atomic energy plants. E. M. Forster's world is indeed almost upon us.

Yet these developments will not necessarily mean an over-all reduction of transportation. I see a great reduction of transportation *for work,* but increased transportation for pleasure. So you need not sell your airline stock right away.

A result of this will be that vast uninhabited areas of the Earth could be opened up, because people will have far greater freedom to choose where they will live. I only hope that the world's remaining virgin areas are not spoiled in the process. Recently, when flying over the Grand Canyon, I suddenly thought, "My word! *This* is the prize real estate of the twenty-first century!" Today there's nobody there except a few mules and tourists. I wish it could stay that way, but I'm not very hopeful.

These trends will inevitably accelerate the disintegration of the cities, whose historical function is now passing. Cities will go on growing, of course, like dinosaurs—for the same reasons, and with the same results. I can even see the time when only the uneducated and criminal elements are left in the cities; the wars of 2001 may be internal military operations against the decaying concrete jungles. Watching the TV news, I wonder if the preliminary skirmishes may not already have started.

When people begin to live in strange, remote places, it will be necessary to develop the autonomous or entirely self-contained household. I am indebted to Professor Buckminster Fuller for some of these ideas, which he gave me one morning at breakfast. Bucky thinks that, as a result of the space program, we will develop techniques to reprocess all waste materials so that nothing is lost. Once the research is done, such "closed ecologies" will be available for general use. Single homes, or at least small communities, will then be almost independent of outside supplies for such basics as food and water. They will be able to make everything they need.

This leads to another concept of Bucky's, which one might call the mobile town. When you take one of his famous geodesic domes and make it very large—a mile or so in diameter—the air inside weighs so much more than the dome and its contents that a rise in temperature of a few degrees could make the whole thing take off like a hot-air balloon. So why not go south in the winter and north in the summer—without leaving home?

Now I would like to discuss environment, which is very much a function of transportation and communication. But it is also a function of population. As everybody knows, we are now in a population explosion—but one characteristic of explosions is that they eventually stop and then the bits start to fall down. Probably around the turn of the century this particular explosion will be controlled and the world population may be shrinking again. Although this doesn't help us much, because we'll have to face a population of six billion or so in a couple of decades, it's interesting to speculate about the ultimate figure. I see no reason for more than a very few million people on the planet Earth. Once a population is controlled, one can aim at any level one likes, and reach it in a few centuries.

Nevertheless, even with a six billion population there may be more room than is generally imagined today. By the twenty-first century, agriculture will be on the way out. It's a ridiculous process: a whole acre is needed to feed one person, because growing plants are extremely inefficient devices for trapping sunlight. And when animals feed on the plants, that introduces another 90 per cent of loss. If we could develop a biological system working at a mere 5 per cent efficiency—today's solar cells can double that—it would require twenty square feet, not one acre, to feed one person. The roof of the average house intercepts more than enough energy to feed its occupants!

Food production is the last major industry to yield to technology.

Only now are we doing something about it, probably too little and too late.

One promising field of research is the production of proteins from petroleum by microbiological conversion. (Which sounds most unappetizing—but we do use microbes to make wine!) This process gives high-quality proteins, some of them better balanced for human consumption than natural vegetable proteins. It would take only 3 per cent of today's petroleum output to provide the total protein needs of the entire human race.

It's strange to think that the great grain fields of the West may be gone in another lifetime. . . . This will be an economic as well as an aesthetic loss. Agriculture has given man some of his most satisfying environments—the terraced paddy fields of Asia, the English shires, the neat farms of the Pennsylvania Dutch. Fuzzy-minded romantics often talk of such landscapes as natural, but of course this is complete nonsense. Farms are among the most beautiful machines that man has ever built; and like all machines they will one day be obsolete.

With the exception of luxury items—and the Russians, I've heard, have already started to export synthetic caviar!—most foods will be factory-made in the next century. This will free vast areas of agricultural land for other purposes—living, parks, recreation, hunting—above all, for *wilderness.*

The new societies will need all this new space. With the advent of automation, there will be a great decrease in the old categories of work. In his book *The Environment Game* the British science writer Nigel Calder points out that work is an *invention,* due to the invention of agriculture. Work may be defined as something that doesn't come naturally. Primitive hunters lived tough, hard lives but they didn't work—that's why people still go hunting for recreation. Now, says Calder, with the coming of automation, we must disinvent work—which may not be easy.

However, there are two factors which may help the process. The first will be the arrival of ultraintelligent machines. Today's computers, despite their astonishing powers, are merely mechanical morons: they can't think. But if anyone tells you that they'll never think, that only proves that some humans can't. They'll be thinking all right, by the turn of the century.

The second factor that will lead to the decline of work is the restoration of slavery—I hasten to add, in a morally unobjectionable form. This will be a by-product of the next great technological breakthrough, applied molecular biology.

It's a surprising and indeed mortifying thought that man has acquired no new domestic animals since Neolithic times. The coming science of genetic engineering, together with psychological programming techniques, will give us quasi-intelligent animal servants—probably, though not necessarily, of Simian stock. They will be at least as competent as much of the labor you can hire today through the Yellow Pages—and a lot less trouble, until they become unionized.

These developments will obviously produce vast social problems, but may also solve others by allowing man to revert to ways of life more fitting to his nature. Some aspects of that nature are hardly flattering, and many of our present troubles arise from ignorance of that fact, or failure to admit it. Thanks to the work of Dart, Broom, and Leakey on our African ancestors, and Lorenz's studies of aggression, we are beginning to understand something of our heritage.

Ladies and gentlemen, we are carnivorous predators—the deadliest that this world has ever seen. We need new hunting grounds, psychologically and emotionally if no longer literally. And we are fortunate that technology has now given them to us, in space and in the sea.

These two new frontiers are complementary; we have to develop them both. For the near future the sea is probably much more important than space; in the distant one, space will be more important than the sea. But there is no real conflict.

I have outlined some of the uses of the sea in a number of books, and can merely list them now. Food production, of course, will be of major importance for a long time to come. I'm particularly fascinated by whale ranching and have written a novel (*The Deep Range*) about it. In another book, *Dolphin Island,* I described training—and swimming with—those deadly sea wolves the killer whales, whom I proposed using as oceanic "sheep dogs." Imagine my astonishment recently at seeing Hugh Downs, on the *Today* show, riding on the back of a killer whale! It's tough being a prophet these days. . . .

As a source of raw materials the sea seems inexhaustible. Any element you care to mention is there, in solution or lying on the sea bed. We will also be forced to use it for more and more of our water supply, through desalinization techniques. But perhaps it will not always be necessary to remove the salt: I would like to mention Hugo Boyko's fascinating work in Israel ("Salt-Water Agriculture"—*Scientific American,* March 1967) on irrigating crops directly with sea water. Wherever possible, it is always best to let nature do the work.

Much of the sea is a desert, because the chemicals of life (particularly the phosphates) lie trapped thousands of feet down on the ocean bed, far beyond the reach of sunlight. In *The Deep Range* I suggested bringing them to the surface, using the heat of submerged nuclear reactors to start convection currents. Something like this occurs naturally every spring in the Antarctic, where upwelling currents produce an explosion of vegetable—and hence animal—life. (This annual rebirth is beautifully described in Rachel Carson's *The Sea Around Us.*) If we could trigger the same process in the tropical ocean, the results might be spectacular.

I'm sorry to leave the sea so hastily, but space is a lot bigger and I must spend more time on that.

Our current ideas of space and its potentialities are badly distorted by the primitive nature of our techniques. To prove this, here is a statistic that will surprise you.

The amount of energy needed to lift a man to the Moon is about 1000 kilowatt-hours—and that costs only ten to twenty dollars! The difference of nine zeros between this and the Apollo budget is a measure of our present incompetence. Ultimately, there's no reason why space travel should be, in terms of future incomes, much more expensive than jet flight today.

Moreover, space is a benign—or at least a neutral—environment. It's not ferociously hostile like the ocean deeps or the gale-swept Antarctic. Space communities will be established first on the Moon, then on Mars, and later on other worlds. But much closer to the Earth, orbital space stations of many kinds will be in wide use by the year 2000.

In May 1967 I was in Dallas to attend the first conference on the *commercial* uses of space—including tourism. Barron Hilton gave a talk on the Hilton Orbiter Hotel, which he hopes to see in his lifetime. Space tourism is going to be a major industry in the twenty-first century.

Another tremendously important use of space stations will be for medical research; one paper given at Dallas discussed the engineering problem of a hospital in orbit.

Which brings a poignant memory to mind. The last letter I ever received from that great scientist Professor J. B. S. Haldane * was written when he was dying of cancer and in considerable pain from his operations. In it, he said what a boon the weightless environment of a space hospital would be to patients like himself—not to mention burn victims, sufferers from heart complaints, and those afflicted with muscle diseases. I am convinced that research in space will open up unguessed regions of medical knowledge and give us a vast range of new therapies. So I get pretty mad when I hear ignorant but well-intentioned people ask, "Why not spend the space budget on something useful—like cancer research?" When we *do* find a cancer cure, part of the basic knowledge will have come from space. And ultimately we will find even more important secrets there: perhaps, someday, a cure for death itself. . . .

Now, I've been very well behaved and have stuck so far to technological projects of modest scope. So I trust you'll let me use the last five minutes to mention some rather far-out ideas, just to stretch your minds—not, I hope, to the breaking point.

---

* See "Haldane and Space," page 230.

How would you like to abolish night? This idea has already been mooted in connection with the Vietnam War. It's theoretically possible to orbit giant mirrors in space, to hover over the Equator and to reflect sunlight to any spot on Earth. And as they need only be made of mylar film coated with a few atoms' thickness of aluminum, they would be extremely light even if they were miles on a side. It would be technically feasible to erect such mirrors using the Saturn 5 launch vehicles now under development.

The economic consequences of this could be very great, particularly to the electric lighting industry! There would also be some undesirable side effects—to growing crops and romantic couples, for example. Crime, however, would be greatly reduced, and New York need never again worry about blackouts.

The next project I've christened the Synchronous Skyscraper. With existing techniques, if you're clever enough and use the best materials, you could build structures between five and ten miles high before they collapsed. But would you believe a structure twenty-two thousand *miles* high?

A year ago Professor John Isaacs and his students at La Jolla published a letter in *Science,* pointing out that if one started from the synchronous orbit twenty-two thousand miles up, one could lower cables all the way to the Earth's surface. It would be possible to build an "elevator to the stars" and to use the cable to send payloads into space. This is a really fantastic idea: now the Russians claim to have thought of it first—which proves it *must* be valid. . . .

Finally, I would like to mention some stimulating speculations due to Freeman Dyson of Princeton's Institute for Advanced Study. Dr. Dyson started to think big while working on the classified Orion project, an investigation of very large spaceships in the ten-thousand- or fifty-thousand-ton class, driven by atomic-bomb explosions. This led him to the still more grandiose concept of "astronomical architecture."

Dr. Dyson points out that if a technological society grows at a steady rate, in a few centuries it increases its ability to control mass and power by factors of millions and even billions. So eventually it must control all the resources not only of its planet but of its solar system. In a fairly short time—historically speaking—it might need so much power that it would be forced to enclose its sun to trap all the available energy.

But even that might be only a beginning, as this quote from Dyson's "The Search for Extraterrestrial Technology" shows:

If we assume a technology with a strong drive to expand, it will move from star to star in times at most of the order of 1000 years. It will spread from one end of a galaxy to another in ten million years, still a short time by astronomical standards. We are therefore confronted with a new order of questions. It is not enough to ask: what does a star look like when technology has taken it over? We must ask: what does a galaxy look like when technology has taken it over?

I have the feeling that if an expanding technology had ever really got loose in our galaxy, the effects of it would be glaringly obvious. Starlight instead of wastefully shining all over the galaxy would be carefully dammed and regulated. In fact, to search for evidence of technological activity in the galaxy might be like searching for evidence of technological activity on Manhattan Island. Nothing like a complete technological take-over has occurred in our galaxy. And yet the logic of my argument convinces me that, if there were a large number of technological societies in existence, one of them would probably have carried out such a take-over.

So, ladies and gentlemen, you will be happy to know that your profession still has plenty of room for expansion.

Beyond Babel

The Century of the Communications Satellite

15

There is no longer any need to argue that the communications satellite is ultimately going to have a profound effect upon society; the events of the last ten years have established this beyond question. Nevertheless, it is possible that even now we have only the faintest understanding of its ultimate impact upon our world. The main purpose of this address is to explore some of the further vistas which may be overlooked in our concern with more immediate problems.

Now, I am well aware that the main goal of this conference is to deal with such problems, many of which are so complex that those who have to solve them may be justifiably impatient with starry-eyed prophets gazing twenty, thirty, or forty years ahead. It is necessary to keep a sense of proportion: at the same time we should never forget the goals toward which we are ultimately heading—even if those goals are still vague and ill defined.

But before I attempt to outline the future, I would like to deal with some possible criticisms. There are those who have argued that communications satellites (hereafter referred to as "comsats")

Address to UNESCO Space Communications Conference, Paris, December 8, 1969.

represent only an extension of existing communications devices and that society can therefore absorb them without too great an upheaval.

This is completely untrue.

I am reminded rather strongly of the frequent assertions by elderly generals immediately after August 1945 that nothing had really changed in warfare because the device which destroyed Hiroshima was "just another bomb."

Some inventions represent a kind of technological quantum jump which causes a major restructuring of society. In our century, the automobile is perhaps the most notable example of this. It is characteristic of such inventions that even when they are already in existence, it is a considerable time before anyone appreciates the changes they will bring. To demonstrate this, I would like to quote two examples—one genuine, one slightly fictitious.

For the first I am indebted to the Honorable Anthony Wedgwood Benn, now U.K. Minister of Technology, who passed it on to me when he was Postmaster General. I am speaking from memory, so I do not guarantee the exact accuracy of the quotation.

Soon after Mr. Edison invented the electric light, there was an alarming decline in the Stock Exchange quotations for the gas companies. A Parliamentary Commission was therefore called in England, which heard expert witnesses on the subject; I feel confident that many of these assured the gas manufacturers that nothing further would be heard of this impractical device.

One of the witnesses called was the chief engineer of the Post Office, Sir William Preece, an able man who in later years was to back Marconi in his early wireless experiments. Somebody asked Sir William if he had any comments to make on the latest American invention—the telephone. To this, the chief engineer of the Post Office made the remarkable reply, "No sir. The Americans have need of the telephone—but we do not. We have plenty of messenger boys."

Obviously, Sir William was completely unable to imagine that the time would come when the telephone would dominate society, commerce, and industry, and that almost every home would possess one. The telephone, as it turned out, was to be slightly more than a substitute for messenger boys.

The second example is due to my friend Jean d'Arcy, who is familiar to all of you. He has reported to me the deliberations of a slightly earlier scientific committee, set up in the Middle Ages to discuss whether it was worth developing Mr. Gutenberg's ingenious invention, the printing press. After lengthy deliberations, this committee decided not to allocate further funds, for reasons that I am sure you will agree are extremely logical, and which may strike some responsive chords. The printing press, it was agreed, was a clever idea, but it could have no large-scale application. There would never be any big demand for books—for the simple reason that only a microscopic fraction of the population could read.

If anyone thinks that I am belaboring the obvious, I would like him to ask himself, in all honesty, whether he would have dared to predict the ultimate impact of the printing press and the telephone when they were invented. I believe that in the long run the impact of the communications satellite will be even more spectacular. Moreover, the run may not be as long as we think, and at this point I would like to assert my dubious claim to being a somewhat conservative prophet.

Until very recently I was under the impression that I first advanced the concept of the synchronous comsat in the well-known paper published in *Wireless World* for October 1945. A few months ago, to my great surprise, some friends in the Ceylon Broadcasting Corporation unearthed a letter of mine published in that same journal for February 1945, which I had completely forgotten. It suggested that V-2 rockets should be used for ionospheric research, but the concluding paragraphs described the synchronous communications satellite network and contained the now rather comical phrase: "a possibility of the more remote future—perhaps half a century ahead." I was bravely risking ridicule, predicting communications satellites *by 1995!*

This is the reverse of the usual tendency, which has often been pointed out, for technological forecasts to be overoptimistic in the short run but overpessimistic in the long run. The reason for this is really rather simple. The human mind tends to extrapolate in a linear manner, whereas progress is exponential. The exponential curve rises slowly at first and then climbs rapidly, until eventually it cuts across the straight-line slope and goes soaring beyond it. Unfortunately, it is never possible to predict whether the crossover point will be five, ten, or twenty years ahead.

However, I believe that everything I am about to discuss will be technically possible well before the end of this century. The rate of progress will be limited by economic and political factors, not technological ones. When a new invention has a sufficiently great public appeal, the world insists on having it. Look at the speed with which the transistor revolution occurred. Yet what we now see on the technological horizon are devices with far greater potential, and human appeal, even than the ubiquitous transistor radio.

It must also be remembered that our ideas concerning the future of space technology are still limited by the present primitive state of the art. All of today's launch vehicles are expendable—single-shot devices which can perform only one mission and are then discarded. It has been recognized for many years that space exploration, and space *exploitation,* will be practical only when the same launch vehicle can be flown over and over again, like conventional aircraft. The development of the reusable launch vehicle—the so-called "space shuttle"—will be the most urgent problem of the space engineers in the 1970s.

It is confidently believed that such vehicles will be operating by the end of the decade. When they do, their impact on astronautics will be comparable to that of the famous DC-3 on aeronautics. The cost of putting payloads—and men—into space will decrease from thousands, to hundreds, and then to tens of dollars per pound. This will make possible the development of multipurpose manned space stations, as well as the deployment of very large and complex unmanned satellites which it would be quite impractical to launch from Earth in a single vehicle.

It must also be remembered that comsats are only one of a very large range of applications satellites; they may not even be the most important. The Earth Resources satellites will enormously advance our knowledge of this planet's capabilities and the ways in which we may exploit them. The time is going to come when farmers, fishermen, public utility companies, and departments of agriculture and forestry will find it impossible to imagine how they ever operated in the days before they had spaceborne sensors continually scanning the planet.

The economic value of meteorological satellites—and their potential for saving life—has already been demonstrated. Another very important use of satellites, which has not yet begun but will have an economic value of billions of dollars a year, is air-traffic control. It appears possible that the *only* real solution to the problem of air congestion and the mounting risk of collisions may be through navigational satellites that can track every aircraft in the sky.

All these multitudinous uses of space, although they will compete with comsats to some extent for the use of the available spectrum, will help to reduce the cost of their development and maintenance. The establishment of check-out and servicing facilities in space may therefore be economically feasible several years earlier than would be the case if the only space application was the communications satellite. We see only one part of the picture if we focus our attention too closely on this single use of orbital facilities, and forget the synergistic effect of the others.

In dealing with telecommunications problems it is convenient—and often indeed essential—to divide the subject according to the type of transmission and equipment used. Thus we talk about radios, telephones, television sets, data networks, and facsimile systems as though they were all quite separate things.

But this, of course, is a completely artificial distinction; to the communications satellite—which simply handles trains of electric impulses—they are all the same. For the purposes of this discussion I am therefore looking at the subject from a different point of view, which may give a better over-all picture. I am lumping all telecommunications devices together and am considering their *total* impact upon four basic units in turn. Those units are the Home, the City, the State, and the World.

The Home

Note that I started with the home, not the family, as the basic human unit. Many people do not live in family groups, but everybody lives in a home. Indeed, in certain societies today the family itself is becoming somewhat nebulous around the edges, and among some younger groups is being replaced by the tribe—of which more anon. But the home will always be with us—as in the sense of Le Courbusier's famous phrase "the machine for living in." It is the components of this machine I would like to look at now.

There was once a time when homes did not have windows. It is difficult for those of us who do not live in caves or tents to imagine such a state of affairs. Yet within a single generation the home in the more developed countries has acquired a new window of incredible, magical power—the television set. What once seemed one of the most expensive luxuries became, in what is historically a twinkling of an eye, one of the basic necessities of life.

The television antenna swaying precariously above the slum dweller's shack is a true sign of our times, and there is profound significance in the fact that during riots and similar disturbances one of the first targets of looters is the color television set in the store window. What the book was to a tiny minority in earlier ages, the television set has now come to be for all the world.

It is true that all too often it is no more than a drug—like its poorer relative, the transistor radio pressed to the ears of the blank-faced noise addicts one sees walking entranced through the city streets. But, of course, it is infinitely more than this, as was so well expressed by Buckminster Fuller when he remarked that ours is the first generation to be reared by three parents.

All future generations will be reared by three parents. As René Maheu remarked recently, this may be one of the real reasons for the generation gap. We now have a discontinuity in human history. For the first time there is a generation that knows more than its parents, and television is at least partly responsible for this state of affairs.

Millions of words have been written on the educational use of television—specifically television programs from communications satellites. But we must not overlook the enormous potential of educational *radio* programs, when high-quality global transmissions become possible. There are some subjects for which vision is essential, others in which it contributes little or nothing. As a television channel takes the spectrum space of several hundred voice channels, it should not be used if it is unnecessary. However, simple cost-effectiveness studies may be misleading. The hypnotic effect of the screen may be necessary to prevent the student's attention from wandering, even when all the essential information is going into his ears.

Anything we can imagine in the way of educational television and radio can be done. As I have already remarked, the limitations are not technical but economic and political. And as for economic limitations, the cost of a truly global satellite educational system, broadcasting into all countries, would be quite trivial compared with the long-term benefits it could bring.

Let me indulge in a little fantasy. Some of the studies of broadcast educational comsats—let us call them edsats—to developing countries indicate that the cost of the hardware may be of the order of $1 per pupil per year.

I suppose there are about a billion children of school age on this planet, but the number of people who require education must be much higher than this, perhaps two billion. As I am only concerned with establishing orders of magnitude, the precise figures don't matter. But the point is that, for the cost of a few billion dollars a year—i.e., a few per cent of the monies spent on armaments—one could provide a global edsat system that could drag this whole planet out of ignorance.

Such a project would seem ideally suited for U.N. supervision, because there are great areas of basic education in which there is no serious disagreement. I do not think that ideological considerations play much part in the teaching of mathematics, chemistry, or biology, at least on the elementary level, though I must admit that some small cults still object to the doctrine that the Earth is round.

The beauty of television, of course, is that to a considerable extent it transcends the language problem. I would like to see the development, by the Walt Disney Studios or some similar organization, of visual educational programs that do not depend on language, but only upon sight, plus sound effects. I feel certain that a great deal can be done in this direction, and it is essential that such research be initiated as soon as possible, because it may take much longer to develop appropriate programs than to develop the equipment to transmit and receive them.

Even the addition of language does not pose too great a problem, since this requires only a fraction of the band width of the vision signal. And sooner or later we must achieve a world in which every human being can communicate directly with every other, because all men will speak, or at least understand, a handful of basic languages. The children of the future are going to learn several languages from that third parent in the corner of the living room.

Perhaps, if we look further ahead, a time is coming when any student or scholar anywhere on Earth will be able to tune in to a course in any subject that interests him, at any level of difficulty he desires. Thousands of educational programs will be broadcast simultaneously on different frequencies, so that any individual will be able to proceed at his own rate and at his own convenience, through the subject of his choice.

This could result in an enormous increase in the efficiency of the educational process. Today, every student is geared to a relatively inflexible curriculum. He has to attend classes at fixed times, which very often may not be convenient. The opening up of the electromagnetic spectrum made possible by comsats will represent as great a boon to scholars and students as did the advent of the printing press itself.

The great challenge of the decade to come is freedom from hunger. Yet starvation of the mind will one day be regarded as an evil no less great than starvation of the body. All men deserve to be educated to the limit of their capabilities. If this opportunity is denied them, basic human rights are violated.

This is why the forthcoming experimental use of direct broadcast edsats in India in 1974 is of such interest and importance. We should wish it every success, for even if it is only a primitive prototype, it may herald the global educational system of the future.

If I have spent so much time on this subject, it is because nothing is more important than education. H. G. Wells once remarked that future history would be a race between education and catastrophe. We are nearing the end of that race, and the outcome is still in doubt; hence the importance of any tool, any device, that can improve the odds.

The City

It is obvious that one of the results of the developments we have been discussing will be a breakdown of the barrier between home and school, or home and university—for in a sense the whole world may become one academy of learning. But this is only one aspect of an even wider revolution, because the new communications devices will also break down the barrier between home and place of work. During the next decade we will see coming into the home a general-purpose communications console comprising TV screen, camera, microphone, computer keyboard, and hard-copy readout device. Through this, anyone will be able to be in touch with any other person similarly equipped. As a result, for an ever-increasing number of people—in fact, virtually everyone of the executive level and above—almost all travel for business will become unnecessary.

Recently, a limited number of the executives of the Westinghouse Corporation who were provided with primitive forerunners of this device promptly found that their traveling decreased by 20 per cent.

This, I am convinced, is how we are going to solve the traffic problem—and thus, indirectly, the problem of air pollution. More and more, the slogan of the future will be, "Don't Commute— Communicate." Moreover, this development will make possible— and even accelerate—another fundamental trend of the future.

It usually takes a genius to see the obvious, and once again I am indebted to Buckminster Fuller for the following ideas. One of the most important consequences of today's space research will be the development of life-support and, above all, *food-regeneration* systems for long-duration voyages and for the establishment of bases on the Moon and planets. It is going to cost billions of dollars to develop these techniques, but when they are perfected they will be available to everyone.

This means that we will be able to establish self-contained communities *quite independent of agriculture,* anywhere on this planet that we wish. Perhaps one day even individual homes may become autonomous—closed ecological systems producing all their food and other basic requirements indefinitely.

This development, coupled with the communications explosion, means a total change in the structure of society. But because of the inertia of human institutions, and the gigantic capital investments involved, it may take a century or more for the trend to come to its inevitable conclusion. That conclusion is the death of the city.

We all know that our cities are obsolete, and much effort is now going into patching them up so that they work after some fashion, like thirty-year-old automobiles held together with string and wire. But we must recognize that in the age that is coming the city—

except for certain limited applications—is no longer necessary.

The nightmare of overcrowding and traffic jams which we now endure is going to get worse, perhaps for our lifetimes. But beyond that is a vision of a world in which man is once again what he should be—a fairly *rare* animal, though in instant communication with all other members of his species. Marshall McLuhan has coined the evocative phrase "the global village" to describe the coming society. I hope "the global village" does not really mean a global suburb, covering the planet from pole to pole.

Luckily, there will be far more space in the world of the future, because the land liberated at the end of the agricultural age—now coming to a close after ten thousand years—will become available for living purposes. I trust that much of it will be allowed to revert to wilderness, and that through this new wilderness will wander the electronic nomads of the centuries ahead.

The State

It is perfectly obvious that the communications revolution will have the most profound influence upon that fairly recent invention the nation-state. I am fond of reminding American audiences that their country was created only a century ago by two inventions. Before those inventions existed it was impossible to have a United States of America. Afterward, it was impossible *not* to have it.

Those inventions, of course, were the railroad and the electric telegraph. Russia and China and in fact all modern states could not possibly exist without them. Whether we like it or not—and certainly many people won't like it—we are seeing the next step in this process. History is repeating itself one turn higher on the spiral. What the railroad and the telegraph did to continental areas a hundred years ago, the jet plane and the communications satellite will soon be doing to the whole world.

Despite the rise of nationalism and the surprising resurgence of minority political and linguistic groups, this process may already have gone further than is generally imagined. We see, particularly among the young, cults and movements that transcend all geographical borders. The so-called Jet Set is perhaps the most obvious example of this transnational culture, but that involves only a small minority. In Europe at least, the Volkswagen and Vespa sets are far more numerous and perhaps far more significant. The young Germans, Frenchmen, and Italians are already linked together by a common communications network and are impatient with the naïve and simple-minded nationalism of their parents, which has brought so much misery to the world.

What we are now doing—whether we like it or not—indeed whether we *wish* to or not—is laying the foundation of the first global society. Whether the final planetary authority will be an analogue of the federal systems now existing in the United States or the U.S.S.R. I do not know. I suspect that, without any deliberate planning, such organizations as the world meteorological and earth resources satellite systems and the world communications satellite system (of which Intelsat is the precursor) will eventually transcend their individual components. At some time during the next century they will discover, to their great surprise, that they are really running the world.

There are many who will regard these possibilities with alarm or distaste and may even attempt to prevent their fulfillment. I would remind them of the story of Canute, the wise king of England, Denmark, and Norway, who had his throne set upon the seashore so he could demonstrate to his foolish courtiers that even the king could not command the incoming tide.

The wave of the future is now rising before us. Gentlemen, do not attempt to hold it back. Wisdom lies in recognizing the inevitable—and cooperating with it. In the world that is coming, the great powers are not great enough.

The World

Finally, let us look at our whole world—as we have already done through the eyes of our Moon-bound cameras. I have made it obvious that it will be essentially one world, though I am not foolish or optimistic enough to imagine that it will be free from violence and even war. But more and more it will be recognized that all terrestrial violence is the concern of the police—and of *no one else*.

And there is another factor that will accelerate the unification of the world. Within another lifetime this will not be the only world, and that fact will have a profound psychological impact upon all humanity. We have seen in the *annus mirabilis* of 1969 the imprint of man's first footstep on the Moon. Before the end of this century, we will experience the only other event of comparable significance in the foreseeable future.

Before I tell you what it is, ask yourselves what you would have thought of the Moon landing thirty years ago. Well, before another thirty have passed, we will see its inevitable successor—the birth of the first human child on another world, and the beginning of the *real* colonization of space. When there are men who do not look on Earth as home, then the men of Earth will find themselves drawing closer together.

In countless ways this process has already begun. The vast outpouring of pride, transcending all frontiers, during the flight of Apollo 11 was an indication of this. During those momentous days I was privileged to join Walter Cronkite and Commander Walter Schirra in the CBS-TV coverage of the mission. Mr. Cronkite had previously interviewed President Johnson after his retirement, and this fascinating interview disclosed the most remarkable example of the unifying effect of space exploration I have ever encountered. I would like to pass it on to you now.

After the Moon-circling flight of Apollo 8 in December 1968, President Johnson sent every head of state a copy of the famous photograph of the Earth rising beyond the edge of the Moon. To quote Mr. Johnson, "The response I got to that letter and that picture is truly amazing. The leaders of the world wrote me and thanked me for my thoughtfulness, and expressed great admiration and approbation at what we had done in using space for peaceful purposes."

Then, to the great surprise of Walter Cronkite, President Johnson produced, with evident pride, the personal card of President Ho Chi Minh. He remarked, "Even after I'd returned to the ranch in May, there came a response from Hanoi from Ho Chi Minh thanking me for sending him this picture and expressing his appreciation for this act."

I can think of no better example of the way in which space can put our present tribal squabbles in their true perspective.

Here manned space exploration and the unmanned applications satellites reinforce one another. And it is here that the communications satellites can do their greatest service for mankind. For we are now about to turn back the clock, to the moment in time when the human race was divided.

Whether or not one takes it literally, the myth of the Tower of Babel has an extraordinary relevance for our age. Before that time, according to the book of Genesis (and indeed according to some anthropologists), the human race spoke with a single tongue. That time will never come again, but the time will come, through the impact of comsats, when there will be two or three world languages that all men will share. Far higher than the misguided architects of the Tower of Babel ever could have imagined—36,000 kilometers above the equator—the rocket and communications engineers are about to undo the curse that was then inflicted upon our ancestors. So let me end by quoting the relevant passage from the eleventh chapter of Genesis, which I think could be a motto for this conference and for our hopes of the future.

And the Lord said: Behold they are one people and they have all one language, and this is only the beginning of what they will do, and nothing that they propose to do now will be impossible for them.

IV. *Frontiers of Science*

More Than Five Senses

16

Many years ago, when I was a boy in the country, I invented a rather unkind trick to play on bats. I had long been fascinated by the way in which these strange flying creatures, when they take to the air soon after dusk, are able to locate and catch insects. Even when it is almost completely dark, they will hunt confidently through the sky, suddenly changing direction and darting straight at some invisible moth or beetle.

I knew how it was done, for I had read that bats sent out streams of high-pitched sounds and listened for the echoes reflected from their prey. In our age of radar, of course, everyone is familiar with this idea, but it seemed somewhat fantastic in 1930. Anyway, I asked myself the question: Can a bat tell the difference between an insect and any other solid object in the sky?

So one evening I went out soon after sunset, with a handful of pebbles, and stationed myself near a tall oak where bats were always to be found at dusk. As soon as one flitted overhead, I tossed a stone into its line of flight—and, sure enough, the bat did a power-dive toward it. Indeed, it crashed into the stone with such a thud that I expected it to be stunned.

Almost every time I repeated the experiment, the same thing happened. If the stone passed anywhere near a bat, the creature would make a sharp turn and dive straight at it. Judging by the number of collisions, it was clear that the bat's "radar" could not easily distinguish between insects and stones. This did not surprise me; after all, what sensible bat would expect to find rocks moving around the sky?

Today, we know that bats are not the only creatures who use sound to navigate or to hunt their prey, often in total darkness. Marine animals such as whales and dolphins have developed the sense of "sound location" to a level which we cannot yet approach, even with our most elaborate electronic devices.

When a dolphin is swimming at night, or in dirty water where its eyes are useless, it continually utters a series of squeaks or whistles. We can hear some of these sounds but only a small part of them; most of the noise made by the dolphins is far too high-pitched to register on human ears. But to the dolphin, these sounds are all-important; as they come echoing back from the sea bed, or from schools of fish, they give it a clear and accurate picture of the world through which it swims. Just as a bat can fly through a completely darkened room crisscrossed with wires without hitting one of them, so a dolphin can swim at speed through murky water full of obstacles, avoiding them all.

Yet just as I fooled the bats, sometimes the sea can fool dolphins and whales. From time to time, great schools of these creatures run aground on shallow beaches, become stranded, and die miserably between land and water. This has long been a puzzle to naturalists, and one theory is that a gently sloping beach may not reflect any echoes back to the approaching animals; in certain conditions, it simply absorbs the sound. And so, perceiving no echoes, the poor whales and dolphins continue to swim forward, quite confident that they are heading for the open sea—and discover their mistake too late to do anything about it.

The sound-locating sense of bats, whales, and dolphins is one that we can all appreciate because we share it to some extent. The blind man tapping his stick on the pavement and being warned of obstacles by the pattern of echoes coming back to his ears is doing just the same as the bats and the dolphins, though he cannot do it anything like so well. And in everyday life, all of us—blind or otherwise—use sound for location more often than we suspect.

I once had a dramatic proof of this when I was playing table tennis under a corrugated-iron roof in a tropical rainstorm. The noise was terrific, and my game went to pieces immediately. For the first time, I realized how much I had relied upon the click of the ball upon bat or table to locate it; I shall be rather surprised to find if there are any really good *deaf* table-tennis players. Yet—almost unbelievably—I once saw a blind man acting as a referee at this game; he called every point without hesitation and never missed a fault. It was a wonderful example of what the human ear can do when it is properly trained.

All fish possess a sense organ which we only vaguely understand, because we have nothing like it. It is a thin, irregular line running from head to tail on either side of the fish, called the "lateral line"; apparently it detects changing pressure waves in the water, but this bald statement gives only a faint idea of its capabilities. The first time I saw it in action, I could hardly believe my eyes.

A friend with a large collection of tropical fish was showing me his hundreds of tanks, and in one of them a school of tiny fish was darting back and forth in a restless cloud. Each time it came to the end of the tank, it turned and reversed itself about half an inch from the glass—always at the same distance, just as if it had run into an invisible barrier. I was interested but not particularly impressed, until my friend told me that these little fish were completely blind. Yet on every circuit of the tank they stopped and turned just an instant before they would have charged into the glass walls. How did they do it?

It was not, as in the case of the dolphins and whales, echo location, or sonar, for their feat did not depend on sound. Every fish, when it swims through the water, produces a kind of bow wave like the one you see moving ahead of a boat—though the underwater bow wave is not an up-and-down movement, of course, but a change of pressure. The fish's lateral line can detect this wave; when it nears an obstacle, the wave is distorted by the obstruction ahead of it, and so the fish knows that there is something approaching. It can also spot the pressure waves produced by other fish moving through the water around it—and so it can hunt for food by *feeling,* over its whole body, the currents and vibrations of its liquid world. The vital importance of the lateral line to the fish is proved by the fact that this strange organ is most highly developed in the nightmarish little monsters—all teeth and jaws—that live miles down in the oceans, where no light ever penetrates. In a world where eyes are useless, they must rely on their lateral lines to tell them when to feed—and when to flee.

A good many years ago there was a song that asked the question "Would you rather be a fish?" Scientifically, that's not an easy question to answer, because nobody knows what it would be like to have a lateral line! Perhaps you can get some faint idea of the world of the fish if you stand outdoors on a windy day, wearing no shirt, and with your eyes closed. You can feel the gusts of wind coming at you from various directions; imagine that those gusts represented objects passing through the air around you. If you run quickly, you can feel your own bow wave over your bare skin. But these tiny air currents can give only the feeblest imitation of the rich world of shifting, meaningful pressures in which the creatures of the deep pass their short and hungry lives.

Some fish have evolved a sense organ even more remarkable than the lateral line; they have developed an *electric* sense. They produce pulses of current, at the rate of a few hundred per second (about five times the frequency of our ordinary house circuits) and so set up an electric field in the water around them. The field is generated at the tail of the fish, and picked up by organs near its head. If its pattern could be seen by our eyes, it would resemble the lines of force around a bar magnet, which becomes visible when iron filings are sprinkled over it.

Just as the field around a magnet is warped or distorted if another piece of iron is placed near it, so the field around the electric fish is distorted by the presence of an obstacle in the water. By sensing the changes in the field it produces, the fish can hunt its food and avoid collisions in the muddy waters where it lives.

Please note that this is *not* an echo-sounding system like that used by the bats and dolphins, even though short pulses are involved. (It could work with D.C., but the fish finds it more convenient to use A.C.!) The electric sense is something much more complicated and much less understandable to us than sonar, because we have nothing like it at all.

Although only a very few fish have been definitely proved to possess this peculiar sense, most of them seem to have it in a partly developed form. It has been known for a long time that fish are quite sensitive to electric fields, and this is the basis of the most scientific form of fishing that has yet been discovered. By lowering metal plates into the sea, and charging them to the correct voltage, fish can be compelled to swim into the nets, or even into a pipe through which they can be pumped into a ship! Unfortunately, because sea water is a good conductor and so tends to short-circuit the electric field, this method of fishing has a limited range and requires a considerable amount of electrical power. It works much better in fresh water, which is a rather poor conductor.

Certain fish, as is well known, have gone beyond electric senses and have developed something still more astonishing—electric *weapons*. The discharges produced by electric rays and eels are so powerful that they can stun a man and can probably kill any other fish; there can be few more effective "secret weapons" in the sea. I was once about to spear an electric ray when I recognized it— just in time! The world of electric images and sensations through which these creatures move, and in which they launch their silent thunderbolts at their enemies, is certainly quite beyond our imagination or full understanding.

Human beings cannot detect electric fields; there has never been any reason for them to do so. Our eyes—in daylight, at least— probably do a much better job than the sonic, electric, and pressure senses that the creatures of the sea have been forced to develop.

Perhaps if we lived in a world of perpetual darkness, we might have evolved such senses, or even stranger ones.

It is true that we often feel uncomfortable before a thunderstorm, when there are strong electrical fields in the air. But this sensation is almost certainly due to other causes, such as humidity and heat—not electricity. Yet Nature is full of surprises; perhaps hidden somewhere in our bodies there are sense organs that can respond to electrical fields. If there is any truth in all the countless reports of thought-transference (telepathy) and such mysterious abilities as water divining or dowsing, the answer may lie in some unsuspected electrical sense. I do not say that it is at all likely, but I would hate to say that it is impossible.

Whether any animals—including men—are sensitive to *magnetic* fields is a question that scientists have only started to ask quite recently. As far as we are concerned, the answer is almost certainly "No." If you pick up a magnet, it feels exactly like any other piece of iron. Scientists working in radiation laboratories and nuclear energy establishments have often entered the enormously powerful magnetic fields of their cyclotrons, cosmotrons, and other particle accelerators. Most of them have felt nothing at all; a very few have reported slight sensations around metal fillings in their teeth.

Would a magnetic sense be of any value? To migrating birds and animals definitely, for it would give them a built-in compass, so that they could find the north when there was no other way of telling direction. It has often been suggested that homing pigeons navigate in this manner, and attempts have been made to prove this theory by attaching small magnets to pigeons before releasing them. Confused by the new field, it was argued, the poor birds would be unable to find their way. These experiments have never been very conclusive, and it is now believed that birds rely mainly on the sun and stars for their wonderful ability to navigate thousands of miles, often over the empty sea.

Animals can do so many remarkable things—as the examples already given have shown—that there is a great temptation to invent wonderful and mysterious senses to explain their feats. We must remember, however, that to a hypothetical intelligent being who had no eyes and knew nothing about the power of vision, our own ability to observe events at a great distance would seem a miracle. It so happens that we have developed this particular sense to such a high degree that the others have become much less important.

It might have been the other way round. In some animals, the chemical senses of taste and smell have been so enormously developed that they almost play the part of sight. If you have ever owned a dog, you will know that it spends much of its time in a world that you cannot share—a world of exciting, enjoyable, and sometimes frightening smells. A hunting dog can follow an invisible trail for miles, detecting traces of chemicals that must be present in unimaginably small amounts.

We are very seldom conscious of smells (except when they are bad ones) and must undoubtedly miss a great deal of the richness of the natural world. Many years ago, G. K. Chesterton summed this up very neatly in a poem where he put these sarcastic, if ungrammatical, words into the mouth of a dog:

> Even the smell of roses
> Is not what *they* supposes
> And goodness only knowses
> The noselessness of Man!

Once again it may be in the sea, not on land, that the twin senses of smell and taste are most highly developed. Fish (and perhaps dolphins) may be able to tell where they are in the ocean by sampling the water around them; every sea, and every current in the sea, may have a different taste. It is well known that sharks are extremely sensitive to traces of blood in the water; every skin diver knows that a bleeding fish is likely to attract sharks. The many attempts to develop a shark repellent rely on the hope that there may be some substances that taste intolerable or terrifying to these powerful and dangerous creatures. Despite all you may have read to the contrary, no one has yet found a way of discouraging a really hungry shark; the only repellent that sometimes works is a hard bang on the nose, and if matters have come to this, the situation is already pretty desperate.

Incidentally, sharks would soon starve to death if they had to rely entirely on smell in order to find their food. I have watched a hungry shark swim around and around a bleeding fish, bumping its nose on the rocks within inches of it, yet quite unable to find the food that was in full view. Not until the fish moved, and its scales twinkled in the sun, did the shark strike. Smell is a rather vague sort of sense—it does not allow one to pinpoint an object, in the way that sight and hearing do.

It also has two other serious disadvantages; it is very slow-acting, and it often operates only in one direction. The blood from a wounded fish must take several minutes to travel any distance through water, and it will not go upcurrent. It is just as well, therefore, for the shark that it has other means of locating food. Underwater hunters have discovered, time and again, that sharks appear on the scene within seconds of a fish being speared. They must have been attracted by some sound or vibration—perhaps the desperate fluttering of the injured fish—which they can spot by means of the lateral line already mentioned. This gives them their long-range detecting sense; then they close in, and sight and smell take over. That's when it's time to get out of the water.

Of all our senses, the one which, we feel, puts us most closely into direct touch with the real universe is the sense of touch. I have deliberately left that statement in its clumsy form to show how the natural choice of words emphasizes this very point: "the sense which, we *feel,* puts us most closely into *direct touch. . . .*" The long-distance senses of sight and hearing can be easily misled; if we want to be quite sure that an object is really what it seems to be, we reach out and touch it.

Some animals can reach very much farther than we can; they have turned touch into a medium- or even a long-range sense. Cats and many deep-sea fish have done this by the simple trick of growing whiskers or feelers, but they are all beaten by the spider, which sits at the center of a great web hundreds of times larger than its own body, ready to detect anything blundering into it. The spider has, in effect, built an artificial world so that its sense of touch can be extended over a vast area. Looking at it from this point of view makes us realize what a wonderful achievement a spider's web really is; it is much more than a trap—it is a communications network. There is nothing else like it until we get to man and his telephone systems.

To round off this survey, let us leave the familiar (and not so familiar) creatures of our own planet and let our imaginations roam out into space. It is interesting and amusing—and one day it may be very useful—to ask ourselves what strange senses have been developed by the creatures of other worlds, in conditions totally different from those we know on Earth. (Though Earth itself can provide quite a spectacular range, from the depths of the Pacific to the summit of Everest, from boiling lava pools to the howling subzero winds of the Antarctic.)

Perhaps there are life forms, somewhere in the Universe, that can detect radioactivity, which we can do only with instruments such as Geiger counters. Such a sense would not evolve unless there was an urgent practical value for it; one can, for example, imagine a planet with large areas of radioactivity which would be dangerous to approach. We—and all the other animals on this Earth—would walk into such areas without receiving any warning at all. Stretching our imaginations still further, we can conceive of creatures that actually *needed* radioactive elements to keep them alive, and so were compelled to develop senses to detect them. This is fantasy, of course; but the Universe is so fantastic that everything that is at all possible must happen somewhere. An animal that could detect radioactivity is no more astonishing than the fish that "feels" with electricity.

We know many forces and powers that were undreamed of only a few hundred years ago; there must be many more still undiscov-

ered. Our grandfathers were stunned by the news that there were
rays—X rays—that can pass through solid matter, as light passes
through glass. Are there creatures anywhere in the Universe who
can see by X rays?

If they exist at all, they cannot live on planets like ours, for air
absorbs X rays very rapidly, and on Earth a being with X-ray vision
could see only a few feet (so much for Superman!). But on an airless world circling a sun so hot that much of its radiation occurred
in the X-ray band, some kind of X-ray vision would be theoretically
possible. An X-ray "eye" would be a very peculiar organ, because
any lens would be useless. It might be a pinhole camera made of
lead; once again this is pure fantasy, but nature has created
stranger things.

You may care to amuse yourself by inventing still more unlikely
—yet scientifically possible—organs of sense. And to put you in
the right frame of mind for the task I would like to mention a famous painting, called "The Blind Girl," by the Victorian artist Sir
John Millais (1829–1896). It shows a beautiful English landscape
with a thunderstorm in the distance and a glorious rainbow arched
across it. The whole composition is in that detailed, photograph-
ically exact style so unpopular today because it requires hard work
and a brilliant technique.

In the foreground sits a blind girl, unaware of all the beauty
around her. A butterfly has alighted on her shawl, and her little
companion—perhaps her sister—looks at it with wonder. To the
blind girl, both butterfly and rainbow might not exist.

It is a touching picture, and it still moves me although I have
not seen it for twenty years. And it teaches an even deeper lesson
than the one that the artist intended.

We think that we see and hear and touch and taste and smell the
world around us well enough to know it as it really is. Yet compared with bats and dolphins we are as good as deaf; to dogs we
must appear to suffer from a permanent cold in the nose; and our
eyes can see only one narrow band of the whole spectrum of light.
Of electrical, magnetic, or radioactive senses, we have no trace.

The Universe has existed for billions of years, and the human race is very young. There may be creatures among the stars who have evolved all the senses that we can imagine, and many more. They would pity us as we pity Millais's blind girl.

Many years ago an American poet, whose name I have forgotten and would be delighted to rediscover, summed up this thought perfectly, in four lines which express all that I have been trying to say in several thousand words.

> A being who hears me tapping
>   The five-sensed cane of mind
> Amid such greater glories
>   That I am worse than blind.

Read that verse carefully, and ponder its meaning. Once you understand it, the world will never again be quite the same to you.

Things That Can Never Be Done

17

A good many years ago I came across a simple but instructive brain teaser that I'd like to pass on to some fresh victims. Three new houses are waiting for their gas, electricity, and water supplies to be laid on by the utility companies. Unfortunately, the companies are on bad terms with each other; once one of them has dug a trench, it won't allow the others to cross it. The problem, then, is to bring the gas, electricity, and water to each of the three houses by routes *which never intersect.*

I suggest that you now take time off with pencil and paper to see if you can arrive at a solution. It doesn't matter where the houses stand, or how winding and roundabout a route the pipes or cables take; the only requirement is that they must not cross each other.

This problem was given to me by an old gentleman who said he'd spent years trying to solve it. I very soon discovered that no matter where you dug the trenches or laid the connecting lines, you would always end up with one that couldn't be connected without breaking the rules. Moreover, you can move the three houses, and the three utilities, anywhere you like. *Whatever* you do, the problem—although it seems such a simple one—is insoluble.

There are many such problems, and they are often much more interesting than those that *can* be solved. Three very famous ones have come down to us from the Greeks; though they appear almost as simple as the one just described, they have engaged some of the best minds in history for more than two thousand years.

The first—and most celebrated—is that of "squaring the circle." All that you have to do is to work out a geometrical construction, using *only ruler and compass,* by which you can draw a square exactly equal in area to any given circle.

No one can estimate how many millions of man-hours have been spent in vain attempts to find such a construction. After several centuries of fruitless effort, most mathematicians began to suspect that the feat was impossible—but they could not actually prove it. Until less than a hundred years ago, there remained a faint hope that somebody might, someday, discover a way of squaring the circle. Then, in 1882, a German mathematician named Lindemann proved conclusively that it was impossible. After that date, anyone who continued to try was a crackpot. It is true that there are many ways of constructing squares *approximately* equal in area to circles, and some of the constructions are so good that for all practical purposes they are perfect; no eye, and no measuring instrument, would show any error. But they are not *mathematically* exact; there always is an error, which can be calculated, even though it may be far too small to be seen.

The other two famous problems? The second is to divide any given angle into three equal parts; the third is to construct a cube twice the volume of another. All three problems, I might add, can be solved exactly *if you are allowed to use special instruments*— but that is against the rules of the game. To bring in any other instrument besides ruler and compass is cheating—like laying a gas pipe over an electric cable in the problem we started with.

I do not know why squaring the circle has attracted so many cranks for so many centuries, but people who have never heard of Lindemann's proof (and wouldn't understand it if they had) are still claiming to have done what we now know to be impossible. Not long ago, I am sorry to say, a United States Senator read a statement into the *Congressional Record* asserting that one of his constituents had not only squared circles but, for good measure, trisected angles and duplicated cubes! This nonsense is an example of the lack of scientific education among its legislators that now threatens the United States' position in the world. However, we do make some progress; fifty years ago, I am sure, there were Congressmen who believed that the Earth was flat.

Let us leave mathematics for a moment, and consider another famous impossibility that kept inventors busy for a good many centuries. This is the perpetual-motion machine.

Now there is nothing absurd about perpetual motion; everywhere we look—at the planets wheeling around the Sun, or the electrons circling the heart of the atom—we see examples of it. Where there is no friction, as in airless space, an object can keep moving forever.

Although we cannot reproduce this state of affairs on Earth, we can get quite near it. A heavy flywheel, supported magnetically in a vacuum, would continue to spin for many years once it was set in motion. I once saw a small electrical device which has, I believe, been running continuously *on its own power* for more than a hundred years and must certainly be the oldest operating "electric motor" in the world. It was a tiny pendulum swinging back and forth between the two contacts of a battery (a small voltaic cell). When it hit one contact it would charge up, be repelled, swing to the other, discharge, and so on—year after year. Eventually, of course, the battery will run down; but unless the string breaks, the gadget will probably still be ticking away into the twenty-first century.

The old dream of the perpetual-motion enthusiasts was something much more ambitious than this. They wanted to build machines which would not only run forever—*but would do useful work while they ran.* This would certainly be getting something for nothing, but in the days before people understood the principles of science and mechanics, the idea did not seem as absurd as we now know it to be.

Most of the perpetual-motion machines that were envisaged—

and quite a few were actually built—were supposed to be driven by gravity. A favorite design was some kind of wheel with movable weights around the rim; the weights were supposed to pull the wheel down on one side, then slide into positions where they would be lifted back to the top on the other side with the minimum of effort. If more energy could be gained on the downward movement of the weights than was lost in carrying them back to the top, then clearly the machine could go on doing work forever—or at least until it wore out.

Some perpetual-motion machines were so ingenious, and so complicated, that it would require a good deal of hard calculation to show exactly where the design was wrong. However, there is no need to go to the trouble; today we know that the whole idea of such a machine is a complete fallacy. If anyone claimed to have invented a bottle or tank from which an unending supply of liquid could be drawn, we would laugh at him. He might produce elaborate drawings showing a complicated network of pipes and chambers which, he claimed, "multiplied" the fluid, but we wouldn't bother to examine them. We would know, without going into details, that you can draw only a gallon of liquid from a one-gallon tank—and that's that. A "perpetual tank" is so ridiculous that as far as I know even the most cracked of crackpot inventors has never tried to make one. (Though the idea is popular in many fairy stories and myths.)

Now energy is just as real as matter; you can alter it, but you can't create it or destroy it. It cannot be manufactured *out of nothing,* any more than matter can be manufactured. So a machine that produced energy indefinitely is on exactly the same level as a tank that can never be emptied. All the "working models" that have been demonstrated in the past—and at one time perpetual-motion machines were as popular as gold mines for extracting cash from people with more money than sense—were ingenious frauds.

The impossibility of perpetual motion does not rule out machines driven by forces that are unknown today. But the energy will be coming from somewhere else—it won't be created in the machine.

To our great-grandfathers, the engines that drive an atomic submarine would seem to be an example of perpetual motion, since they can produce energy for years from no apparent source of fuel. But, of course, fuel is "burned"—the uranium atoms in the reactor are slowly used up and eventually have to be replaced. Nature never gives something for nothing, and this is the fundamental law that the perpetual-motion seekers failed to understand.

An even more famous, and equally vain, quest in times past was the search for the "Philosopher's Stone"—a substance which would turn base metals such as lead or mercury into gold. The alchemist's goal of "transmutation," as the conversion of one element into another is called, has been achieved in our own age; the Atomic Energy Commission now manufactures, by the ton, elements that never existed in nature. And because we have learned how to transmute atoms, and understand something of the tremendous forces that hold them together, we know exactly why the efforts of the alchemists were bound to fail. Even the most violent *chemical* reactions are millions of times too feeble to disturb the interior of the atom. The alchemists were like safe-crackers trying to cut through armor plate by brushing at it with feather dusters. But let us not scorn their centuries of laborious, messy, and often dangerous toil, for they laid the foundations of chemistry.

"Impossible" is a dangerous word, and we have to be very careful how we use it. So many things have been done which, not long ago, were declared to be impossible that there is now a tendency to go to the other extreme and to declare that *nothing* is impossible. This line of argument is very popular with such cranks as circle-squarers and inventors of perpetual-motion machines; when you attempt to prove that they are wrong, they reply, "Ah! Scientists used to say that we could never fly, or travel faster than sound, or send a rocket to the Moon. And now look what's happened; one day, *I* shall be proved right!"

It's not easy to answer this argument, because it does hold a grain of truth. In the past, many scientists have made fools of themselves by making what are called "negative predictions"—that is, assertions that something could never be done. (If I may be allowed to slip in a commercial, you'll find some examples of these in my book *Profiles of the Future*.)

So we had better say, even though it is rather bad logic, that some things are more impossible than others. The only impossibilities of which we can be *absolutely certain* are in the realm of mathematics. Let me give an almost ridiculously simple example.

Consider the fraction ⅓. If you try to express this in decimal notation, by dividing 3 into 1.00000 . . . you will get the answer 0.33333 . . . This is known as a recurring decimal; it goes on for ever and ever, repeating itself endlessly. You can be quite, quite certain that no matter how far you continue the calculation, each term will be an identical 3. You will *never* come to the end of the line—as you do, for example, when you work out ¼ and arrive at the answer 0.25. So if anyone claimed to have found an *exact* answer for ⅓ in decimals, you could be sure that he was wrong, without even having to think further about it.

As I have said, this example is ridiculously simple, but there are others where the truth is not so obvious, and it remained hidden for centuries. The classic case is our old friend π, the ratio between the circumference of a circle and its diameter.

When we first meet π in elementary maths, we are told that it is approximately equal to 22/7. However, π cannot be expressed *exactly* by any simple fraction, though some of them (355/113, for example) give answers that are close enough for all practical purposes.

For more than two thousand years, mathematicians who enjoyed doing long calculations spent large portions of their lifetimes trying to work out an exact value of π. By the middle of the nineteenth century, it had been calculated to over two hundred places of decimals. In 1873, an Englishman named Shanks worked it out to 707 places. (Alas, he made a mistake in the 528th decimal, so the last 180 figures of his answer were worthless.

For the first couple of thousand years, there seemed a chance that π would eventually come out to an exact value; the patient calculators could cherish the hope that one day they would find themselves confronted with a series of 0000000's, and would know that they had come to the end of the trail. However, in 1882 this was finally shown to be impossible. Though only mathematicians can understand the proof, we can now be absolutely certain that the decimal representing π *never* comes to an end.

During the last few years, giant electronic computers have worked out the value of π to over 10,000 places, performing in a few minutes the calculations that occupied men like Shanks for the better part of their lives. Doubtless the still more powerful computers of the future will do even better; someday we may know the value of π, if we wish, to millions or billions of decimal places, which leads us to the following strange conclusion:

Imagine that one day men build a giant electronic brain, which can perform millions of calculations a second, and set it to work grinding out the value of π. Year after year the numbers come pouring out of the machine; and one day it starts to produce a long string of zeros.

Does this mean that the calculations have ended—that π has finally "come out" exactly? No; it is only the operation of the laws of chance. If you toss pennies long enough, you may get ten heads in a row—or even a hundred, though you will have to make several million million million million million tosses before *that* happens. In the same way, there must be places along the line of numbers in π, stretching for light-year after light-year, where blocks of zeros of *any* length you care to mention will come up. But sooner or later the 0's will end; even if the computer poured out nothing but zeros for year after year, we could still be certain that *eventually* the other digits would start to turn up again. For we know now, beyond all doubt, that the number π is infinite in length. To work out its exact value is something that can *never* be done, no matter how long the Universe may last.

There are other mathematical results, some of them extremely simple, where we cannot be quite so certain of the truth. Perhaps the most famous of these concerns a ridiculously elementary proposition known as Fermat's Last Theorem, after the mathematician Pierre de Fermat, who laid it down in 1637.

Everyone knows that there are pairs of numbers which when squared and added together give another perfect square. Thus if you square and add 3 and 4, you get $9 + 16$ or 25—which is the square of 5. The equation

$$3^2 + 4^2 = 5^2$$

is only one of an infinite number of such relationships involving squares. Another is

$$5^2 + 12^2 = 13^2$$

Well, if this can be done with squares, why can't the same thing be done with cubes, or with still higher powers? Rather surprisingly, no one has ever found such groupings of numbers, and Fermat laid it down as a general law than none existed.

Mathematicians are almost sure that this is true—it has been tested for millions of cases—but they have never been able to *prove* it beyond a shadow of doubt, though they have now been trying to do so for over three hundred years. (What makes this particular case so tantalizing is that Fermat himself claimed to have discovered a proof; unfortunately, he never wrote it down!)

So here we have a statement not quite as certain as the impossibility of working out an exact value of π. We cannot be sure that, somewhere up in the quadrillions or decillions, there aren't two numbers which when raised to some power and added together won't give a third number raised to the same power. (The law has, incidentally, been proved for cubes, if you ever see an equation like

$$2864173^3 + 5481247^3 = 6931687^3$$

you will *know* that it's wrong, without bothering to work it out.)

Fermat's Last Theorem remains what mathematicians call a "conjecture"—something believed to be true but not yet proved. There is great excitement in the world of mathematics when a conjecture is finally proved—or, as sometimes happens, disproved. Sooner or later some mathematician is going to gain immortality by writing "Q.E.D." after Fermat's Last Theorem, but it may take a little time. After all, it was more than two thousand years before the hunt for the last figure in π had to be called off. . . .

There are some things that are impossible by their very definition. They involve paradoxes, or self-contradictions. A good example is the old question "What happens when an irresistible force meets an immovable object?" Clearly, once you admit the possibility of a force that can't be resisted, you deny the existence of an immovable object—and vice versa; so such a meeting can never take place. The story of the chemist who invented a universal solvent, and spent the rest of his life looking for something to keep it in, belongs to this same class of paradoxes.

Outside the realms of logic and mathematics, it is hard to draw an absolutely clear line between the possible and the impossible—to say whether a thing can or cannot be done. More usually, we are concerned with the question "Is it worth doing?" and this is often even more difficult to answer. For example, twenty years ago, it was impossible to fly faster than sound. Today the big question is: will SST's pay for themselves? We won't know the answer to this until about the time that the first men have landed on the Moon.*

Which leads us to what is, perhaps, the most famous "impossibility" of our time. The coming of the Space Age has been so swift, and so sudden, that many people forget the things that were said and written about space flight ("Buck Rogers stuff!") only a few years ago. Before 1945, there were very, very few scientists prepared to admit that space travel would *ever* be possible; many wrote articles "proving" that the whole idea was utterly ridiculous. The distances were too great, the power needed too enormous—and so on and so forth. Some of these articles make very amusing reading today.

* [This was optimistic!]

Yet history is repeating itself; now that we all know that men will soon be traveling around the Solar System, some scientists have foolishly gone on record with the statement that travel to the planets is all very well—but we will *never* be able to reach the million-times-more-distant stars. One distinguished physicist remarked recently, "All this stuff about traveling around the Universe belongs back where it came from, on the cereal box." He seems to have forgotten that, not so long ago, most of the hardware now standing at Cape Kennedy was more or less confined to cereal boxes —so, if anything, this is a recommendation. Travel to the stars is going to be extremely difficult and will involve techniques and inventions not yet discovered, but one day it will be achieved.

Every man will have his own ideas of what is possible, and what will be forever impossible; only time will prove if he is right. Here, to set you thinking, is a list of far-fetched projects that philosophers, writers, mystics, and scientists have speculated about for centuries:

>Immortality
>Invisibility
>Time travel
>Thought transference
>Levitation
>Creation of life

For my part, there is only one of these that I feel certain (well, practically certain!) to be impossible, and that is time travel. At the other extreme, the creation of life seems almost a certainty, in the not-too-distant future. As for the rest—I prefer to sit on the fence. You may have other views, and you may well be right.

Perhaps it is just as well that we cannot always, in advance, distinguish between the possible and the impossible. It would be rather a dull world if everything had a forgone conclusion; there is much truth in the old saying that it is better to travel hopefully than to arrive.

Because of the laws of nature and of logic, there will always be things that can never be done. And sometimes the effort of discovering *why* they cannot be done leads to results far more valuable than the goal originally sought.

If the alchemists had discovered the Philosopher's Stone—well, there would have been a lot of gold around by this time. But they discovered chemistry; so instead of a few million tons of soft yellow metal, we have anesthetics and penicillin and synthetic fibers and dyes and vitamins.

Which would you prefer?

The World We Cannot See

18

Stand out of doors on a bright, sunny day—and close your eyes. At once, the world around you vanishes; apart from any sounds that may reach your ears, you cannot tell that it exists. Even the noises you do hear—the far-off whine of a jet, the cry of a bird, the roar of motor traffic, the low murmur of human voices—have meaning only because your eyes, long ago, showed you how they were made. It is through vision that we gain almost all our knowledge of the world around us. It's no wonder that to anyone born with sight blindness seems the most terrible of all afflictions.

And yet our eyes do not show us everything. They have serious limitations, and even some unmistakable defects. All around us there are things that we cannot see—to which, even with eyes open, we are completely blind. And there are some things that we would not understand, even if we could see them, any more than a bushman from the African jungle would understand the lights of Broadway, were he suddenly dumped in Times Square.

But first, what is vision—how do we see? This question puzzled men for centuries, and some ancient thinkers—the Greeks, for example—came up with very peculiar answers. At one time it was believed that the eye observed the world by shooting out particles of some kind, like a stream of bullets. If this were true, it is rather hard to understand where the Sun comes into the picture; one would think we would see just as well in darkness as in light if the eye did all the work.

Nowadays, we know that at least four things are needed to make vision possible. There must be a source of light, such as the Sun or a lamp; there must be an object that reflects the light into the eye; there must be the eye itself; and there must be a brain that understands the images formed inside the eye. So vision is quite a complicated and roundabout process; you can never be sure that you see the same world as your friend, and you can be *quite* sure that you don't see the same world as your dog.

Because it is the key to the whole process of seeing, let us first consider that living camera, the human eye. It is indeed a camera; it has a lens that changes focus to look at objects at varying distances, and it has an iris (or "stop," to use the photographer's term) that is wide open in dim light, almost closed in bright. It does not, of course, carry a roll of film at the back; instead, the eye has a sensitive screen (the retina) on which the image is formed, and where the signals are produced which pass to the brain. Luckily, the details of this process do not concern us here—though I would like to mention one fascinating fact about it. You know that if you have to take photographs in faint light, you change to a fast film; well, the eye does something similar. The retina contains two different types of cells, one working in bright light, the other in dim. So the eye is like a camera which is loaded both with fast film *and* with slow film; but it takes several minutes to switch from one to the other—which is why you are almost blind at first, when you step from a brightly lit room into darkness.

Everyone who has ever played with a convex lens knows that the image it forms is upside down. The old-time studio photographer, crouched under his black velvet cloth and looking at the picture on the ground-glass screen, had to grow accustomed to seeing the world standing on its head. And since the eye is a camera, it follows that the image it produces on the retina is also upside down!

Please stop for a moment and consider what this means. As your eye scans along this line, the words you are reading flash *upside down* on the retina. At the back of your eye, this "V" really appears as a "Λ"!

Yet, of course, you don't see an inverted world. The image at the back of the eye is only the first step on the way to the brain, and somewhere farther down the line—in the maze of switches and circuits formed by the nerves—the picture is turned the right way up. But this is not an automatic process; it is one that every baby has to learn in the first few months of its existence.

What is more, the process can be *un*learned. Some devoted experimenters, in their attempts to unravel the workings of the eye, have worn special glasses that make everything appear upside down (or, what is almost as confusing, turned from right to left, as if seen in a mirror). After these glasses have been worn continuously for some days or weeks, the brain learns to reinterpret the images it receives, so that the world once more appears normal. One scientist grew so accustomed to his left-to-right view that he was able to ride a motorcycle through city streets—though he had to start adapting all over again when he took off his spectacles and once more saw the world he had known all his life!

I mention these curious facts because they remind us that the eye does not work by itself; it is part of a wonderful and complicated system in which the brain plays an even more important role. But the eye is the window through which the brain receives most of its impressions of the outside world; let us now see how well it does its job.

In some ways, you will be surprised to discover, it does a very bad one. A famous scientist once remarked that if a manufacturer tried to sell *him* an optical instrument as badly designed as the eye, he would send it straight back. But this is not a very fair comparison; an instrument maker has a wide choice of many types of glass when he builds a camera or a telescope, and he can combine them in ways that cancel out optical defects. The eye has to make do with water and jelly; in these circumstances, it is a marvel that it works at all.

As far as the over-all sharpness of the image is concerned, even the cheapest camera is much superior to the human eye. The eye, in fact, only sees clearly in a very small area. When you look at a scene such as a landscape, you may *think* that you see it all, but you don't. Your clear view is limited to a tiny circle directly ahead; everything else is blurred.

You can prove this surprising fact without lifting your eye from this page. If you fix your gaze firmly on any letter in this line— say the third "p" in hippopotamus"—then, as long as you don't cheat by moving your eye, you cannot see the letters more than three or four places away from it. You may *guess* them, but you can't see them sharply, and you can only read the whole word because you let your eyes sweep along it. This "scanning" process is going on all the time, and you are quite unconscious of it; but at any given instant you are clearly aware only of a patch that is not much larger than the full Moon in the night sky. Anything outside this patch is a vague blur, until you turn your attention upon it by moving your eyes.

Even then, the field of human vision is still somewhat limited; we cannot see sideways without moving our heads. Yet there are animals which have almost "wrap-around" vision; what sort of picture of the world must a spider see, from the images produced by its *eight* eyes, all looking in different directions? We have one consolation, however: at their best, our eyes are among the sharpest in the whole animal kingdom. A bee or a spider, looking at this page of print, would be quite unable to make out the individual letters.

But the eyes of insects have other surprising powers which the human eye cannot match, for they can see things which are completely invisible to us. And this leads us to the question of color, and the nature of light itself.

One of the first things we learn in optics is that the so-called white light of the sun is really a mixture of all possible colors, and that colors themselves differ only in being waves of varying length. The waves of red light are about twice as long as the waves of violet light; a wave of yellow light comes approximately in the middle, and is about one fifty-thousandth of an inch long. In an age where everyone is used to tuning radio sets, this idea is easy to grasp; but it was difficult to understand when it was first put forward, three hundred years ago.

Just as there are radio waves too long, and others too short, to be picked up by the ordinary domestic receiver, so there are light rays too long, or too short, to be seen by the eye. The short waves are called ultraviolet, the long waves infrared, and though they fall upon us from all directions, we cannot see them. These are the waves to which we are totally color blind.

What would the world look like if we could see into the ultraviolet or the infrared? Would it be very different from the one we see now? When we ask these questions, we may well be putting ourselves in the same position as a blind man who wants a description of a rainbow. However, we can attempt an answer, because nowadays there are many instruments that do allow us to "see" into these invisible regions.

It is also possible for a man to obtain genuine ultraviolet vision, if he takes rather drastic steps to do so. The retina—which, as you will remember, is the screen at the back of the eye—is quite sensitive to ultraviolet rays, but normally none reach it because they are filtered out by the lens before they can get there. If, however, the eye's natural lens is replaced by one of plastic, through which ultraviolet rays *can* pass freely, then it is possible to see ultraviolet.

This operation is often carried out for people who have lost their lenses as a result of eye injury or disease; such people could read an optician's chart, illuminated by ultraviolet light, from top to bottom, in what to the normal person would be complete darkness! I feel that there should be some practical application of this odd fact, but I cannot think of anybody except a burglar or a spy who would be able to benefit from it.

Many insects have ultraviolet vision; this may be merely an accidental by-product of the way their eyes are constructed, of no particular value to them. On the other hand, it may allow an insect such as a bee to distinguish between two flowers that appear identical to us.

The strange, many-faceted eye of the bee can certainly perform one feat which is quite impossible not only for us to match, but even for us to imagine. To understand what this is, we have to look a little more closely at the nature of light.

We have already said that light consists of waves, but they are more complicated waves than those that move on the surface of the sea. As a water wave advances, it rises and falls; we can say that it vibrates in the up-and-down direction, *and in this direction only*. But a beam of light normally contains waves vibrating up and down, right and left, and at every possible angle in between.

However, in certain circumstances a light beam can be made to behave like the more familiar water wave; it can be made to vibrate almost entirely in one plane. When this happens, it is said to be polarized—a word familiar to the general public thanks to the efforts of a youngster named Edwin Land, who developed "Polaroid" while he was still in college. (And then went on to invent, among many other things, the Polaroid Land camera.) When light passes through a sheet of Polaroid, it can be made to vibrate in a single plane, but to our eyes it still looks just like ordinary light.

To a bee, however, "up-and-down" light appears quite different from "right-and-left" light. This is very useful to it, because on a cloudy day when the sun is invisible, the light filtering down from the sky is partly polarized—and the nature of its polarization gives a clue to the direction of the hidden sun. On such a day, a man may be lost, unable to take bearings from the sun; yet a bee can still navigate happily and find its way back to the hive. Its peculiar eyes can thus act as a kind of sun compass, but what polarized light actually *looks* like to the bee is something that nobody can imagine.

The world of ultraviolet light—of waves just a little shorter than visible light—can be explored quite easily with the camera, because all films are highly sensitive to these waves. It appears much the same as our ordinary, white-light world; if your eyes suddenly started to work in the ultraviolet range, you would not

notice the difference until you began to match colors with somebody possessing normal vision. But if you imagine your eyes "tuning," as it were, like a radio set, further and further into the ultraviolet, something very odd would soon happen.

Even though it was broad daylight and the sun was shining
brightly, the scene around you would become steadily darker.
Then, quite abruptly, it would be black as night.

For the air around us simply will not transmit very short ultraviolet rays; it blocks them as effectively as a sheet of paper blocks
ordinary light. So eyes that worked in the short ultraviolet would
be quite useless here on Earth; there would be nothing for them
to see.

However, this is not true out in space. The Sun produces vast
quantities of very short ultraviolet rays, as well as the even shorter
waves we know as X rays. They are all blocked by the atmosphere
twenty or thirty miles above our heads, luckily for us. If they
reached the surface of the Earth, life as we know it would be impossible. These rays are deadly, and astronauts have to be pro-
tected against them.

On the Moon (and probably on Mars), these very short waves
get down to ground level without hindrance. Of course, no one expects to find any advanced forms of life on the Moon—but if there
are any Lunarians and Martians, they would find ultraviolet-sensi-
tive eyes quite useful. And there may be worlds elsewhere in the
Universe, circling hotter stars than our Sun, whose inhabitants can
see *only* in the ultraviolet and would be completely blind in what
to us is visible light. There are good reasons for thinking that this
is rather unlikely, as these rays are so destructive to any form of
life that we can imagine; but nature has managed some very unlikely things here on this Earth. . . .

Now let us go in the other direction, toward waves of increasing
length—as if we were to move down the keyboard of a piano to
notes of deeper and deeper pitch. The analogy between light and
sound, if not pushed too far, is quite a useful one. Ultraviolet waves
are like sounds too high-pitched to be heard; in the same way

there are sounds too low-pitched to be heard. The light waves too low-pitched to be seen we know as infrared, which simply means "below red."

Though we cannot see infrared rays, we can feel them if they are strong enough. Infrared rays are heat rays; if you hold your hands in front of a hot electric iron, you can tell that it is there even if you cannot see it. It might almost be said that we have infrared "eyes" all over our body, in the heat-sensitive cells of the skin. They cannot form a definite image, but they can tell us when infrared radiation is present. There are some primitive animals that react to light in this simple way; though they cannot really see, they can tell the difference between light and darkness. That is the best that we can do with infrared—and even then it has to be very strong before we can detect it.

It is quite easy to photograph the world of the near infrared, using ordinary cameras and special film, and it turns out to be a somewhat peculiar place. Trees and plants which look dark green in visible light appear very bright in the infrared; indeed, photographs of vegetation look like winter scenes. Leaves and grass seem covered with snow, they are so dazzlingly white.

As is well known, infrared rays penetrate haze and mist (though not cloud) and are, therefore, valuable for aerial photography. Hawks and other birds of prey would find infrared vision very useful; I should be surprised if they did not have it, at least to some extent.

As we move further away from the realm of visible light and explore the behavior of longer and longer waves, a strange thing begins to happen. I can best explain it by describing a simple experiment.

Imagine a blacked-out room with three objects in it—a live mouse, a bunch of flowers, a cube of ice. (An odd selection, I know, but there's method in my madness.) An electric light is shining on them, so all three objects are easily visible.

You switch off the light; what happens? Of course, the objects disappear; there's no light in the room to see them by, so this is just what one would expect. End of Experiment No. 1.

Now imagine that you have an eye that is sensitive to long infrared rays, and repeat the experiment. When the light is shining, you see all three objects very well, for an ordinary electric bulb produces torrents of infrared. (Indeed, it's a much better infrared generator than it is a source of visible light!)

Now switch off the light; the room darkens, of course—*but you can still see*. The walls are a dim background glow, the mouse is a bright blob, the bunch of flowers a fainter one. Only the ice cube seems completely black—but even that, when you examine it closely, can be seen to have a very faint and feeble glow.

What you are seeing is the heat radiation from these bodies. Every object "shines" in the infrared, because it possesses heat— and infrared rays are simply the rays of heat. The mouse looks brightest because it is a small, active animal with a high body temperature. And though you may think of an ice cube as something cold, it is hundreds of degrees hotter than, say, liquid hydrogen or the night side of the Moon.

During the last few years, scientists have developed sensitive infrared detectors which allow us to "see" objects by their own heat. But, as is so often the case, nature has already beaten us to it.

The snakes known as pit vipers (a family that includes the rattlesnakes and moccasins) have small pits on either side of their heads. These are heat-detecting organs—"eyes" that can "see" in the infrared. They allow their owners to hunt at night, in utter darkness, searching for warm-blooded prey by the heat emitted from their bodies. Only during the last few years have we been able to build guided missiles to do the same, homing on the heat of jet exhausts in the sky.

Because the world of the far infrared is really the world of heat patterns, it would appear strikingly different from the familiar world of visible light. Cold objects would appear dark, hot objects white, while the shadings in between would correspond to lukewarm temperatures. This fact has very important industrial, scientific, and military applications, only a few of which I can mention here.

An engineer can observe a piece of machinery through one of the recently developed infrared detectors and can see at once any dangerous hot spots where overheating is taking place. A doctor can examine a patient, and a tumor hidden inside the body may reveal itself by the glow of excess heat it is producing. An orbiting satellite can scan a country and can pinpoint underground factories and secret installations—especially atomic reactors—by the heat they generate.

Strangest of all, there are circumstances in which infrared detectors will allow us to "see" into the past! Imagine a runway, with an aircraft on it waiting to take off. The tremendous heat from the jets warms up the concrete—and many hours later that invisible glow is still there. An infrared survey can reveal how many aircraft have operated from a field by spotting the heat trails they left behind them—just as one can deduce the number of snails that have crawled through a garden in the night, by the glistening tracks that remain next morning.

One would expect infrared vision to be highly developed among creatures living on the planets of cool, red stars—if such planets exist and are inhabited. Infrared "eyes," however, would have serious limitations: they would give rather coarse and fuzzy pictures, for the images they produced could not be sharply focused. A typical infrared heat wave is about a hundred times longer than a wave of visible light, which means that infrared eyes with vision as sharp as ours would have to be a hundred times larger. I would not say that an eye eight feet across is impossible, but it would certainly be inconvenient!

Moving down through the infrared to regions of yet longer wave length, we come once more into familiar territory. We meet first the centimeter-long waves of radar, then the meter-long waves of the so-called short-wave band, and finally the waves of the broadcast band, which are a few hundreds of meters in length. I said that these waves are "familiar," yet we have known and used them only since the beginning of this century, and radar waves ("microwaves") are barely thirty years old.

We have no senses that can detect radio waves; nor, as far as we know, have any animals. There are very good reasons for this.

Until mankind started producing them in quantity, around the 1920s, sources of radio waves were few and far between. A creature that could see by radio waves would have had nothing to see, except in brief flashes during thunderstorms. There would also be a very faint radio glow from the sky, and occasionally from the Sun; that is all.

Radio vision would be even more limited than infrared vision; it could make out details only in extremely large objects. Any radar set demonstrates this, for it has an "eye" (its antenna or scanner) many feet across, yet two objects have to be several yards apart before it can distinguish between them. A radar set has such coarse vision that a few pieces of reflecting metal foil and a large bomber may look identical.

The dim and flickering world of nature's own radio sources can be observed only by the gigantic metal mirrors of our radio telescopes, and it involves objects of astronomical sizes—planets and stars and galaxies, not the everyday things of this Earth. Yet places exist in the Universe where radio waves are more intense than light waves, and under the blinding glare of those strange radio skies, in conditions almost beyond our imagination, creatures may have evolved who can use radio as we use light.

It might be quite easy to communicate with them through our radio transmitters—but they would be wholly unable to "see" our bodies; we would be like ghosts in their world.

As other creatures, perhaps, are ghosts in ours. . . .

Postscript

Since this essay was written a friend of mine, the well-known British telescope maker Horace E. Dall, has demonstrated a scientific use for ultraviolet vision. Being unlucky enough to undergo cataract operations involving the removal of the crystal-line lenses, Mr. Dall turned this misfortune to advantage. He can now see down to 3,300 angstroms (the normal eye cuts off around the violet at 4,000 angstroms) and reports that in the ultraviolet, Mars is barely visible and such bright red stars as Betelgeuse and Aldebaran cannot be seen at all. Even the familiar constellations change their appearance. Thus only two stars are visible in the Great Bear! (See "Visual Astronomy in the Ultra-Violet," *Journal of the British Astronomical Association* 75, No. 5, August 1965.)

I am indebted to Mr. Rostrom, of Evanston, Illinois, for the following information. During World War II, the Office of Strategic Services (OSS) used "brave elderly people" who had undergone cataract operations to pinpoint the flashing ultraviolet signals of its agents on enemy coasts. These were completely invisible to anyone else. (See *Of Spies and Stratagems,* by Stanley Lovell, Prentice-Hall, 1963.)

Things in the Sky

19

During a recent lecture tour of the United States I was astonished (and disturbed) by the continuing extreme interest in "flying saucers." I had been optimistic enough to assume that everyone was as bored with them as I was—but no; they cropped up in at least 50 per cent of the question periods. And although enthusiasm for aerial crockery rises to a sharp peak in the region of California, it is still rampant on both sides of the Atlantic. Indeed, on my last transit through England I recklessly jeopardized my place on future Honours Lists by getting into a brisk argument on the subject with Royalty.

The reason why I don't believe in flying saucers (few of which are saucer-shaped anyway) is that I've seen far too many. And so will any person of normal eyesight during the course of a few years, if he bothers to look at the sky at all.

Perhaps I had better amplify that statement, and it might also be a good idea to replace the emotion-laden term "flying saucer" with the less controversial one, "unidentified flying object" (UFO). The point I wish to make is that the sky contains an almost endless variety of peculiar sights and objects, only a few of which any one person is ever likely to encounter in a lifetime. Yet any averagely observant person is bound to see some of them, and not knowing their explanation may be misled into thinking he's seen something incredible—instead of merely unfamiliar.

Let me give an example that may seem a little farfetched, but which makes my point perfectly. Suppose you are completely ignorant of meteorological phenomena and live in a country where it never rains. Then one day you step out of doors—and there is a huge semicircular arch spanning half the sky. It is so geometrically perfect that you feel it must be artificial, yet it is obviously miles across, and it is beautifully colored in reds, blues, yellows, greens.

Well, if you had never seen one before, what would you make of a rainbow? It no longer creates the slightest surprise, because it is so familiar; and we, unlike our ancestors, have no need to invent supernatural explanations for it. Reason has told us what it is, and there would be many fewer UFOs around today if reason—or even common sense—was in better supply.

To demonstrate this, I'll describe some of the odd sights I've seen in the heavens, all of them during daylight and under conditions of good visibility. The first was over London on a bright Sunday afternoon, more than twenty years ago. It must have been a Sunday, for that was the only time I had for long walks through the city.

Somewhere north of Oxford Street, I came across a group of people staring at the sky. Following their gaze, I was surprised to see two tiny black dots or disks, very close together, at a great but quite unguessable height above the city. Balloons? I asked myself. No—they don't travel in pairs. And these dots were motionless, despite the fact that a strong wind was blowing. I looked at them for a long time without being able to resolve the mystery; then, having nothing better to do, I started to walk in the general direction of the zoo, above which the objects were floating. (This, by the way, is what the detective-story writers call a Misleading Clue; the London Zoological Gardens had nothing to do with the matter.)

Before you read any further, I would like you to make a determined attempt at explaining this incident. And when I give the simple answer, please don't say in disgust, "Is that all there was to it?" Remember Sherlock Holmes's sour remark to Dr. Watson, when that paragon of unsocialized medicine commented on the obviousness of some mystery which Holmes had just solved. Not being a member of the Baker Street Irregulars, I can't quote chapter and verse, but the reprimand ran somewhat in this fashion: "It's always obvious to you, Watson, *after* I've explained it."

Well, the twin disks floating high above London turned out to be not two objects, but one—a box kite at an altitude of at least a mile. It was so high that its shape was quite unrecognizable; the framework could not be seen at all, while the silk-covered ends had lost all squareness and appeared as disks or spheres. Never before or since have I seen a kite at such an altitude; the elderly gentleman who was controlling it from Regent's Park was operating a reel like a big-game fisherman's, and when he finally brought the thing to earth it looked like a half-scale model of the Wright biplane.

If you think that one was too easy, let us move on to number two. This was on the other side of the world—in Brisbane, state capital of Queensland. I was in an office overlooking the city (arguing, if I remember correctly, with a customs inspector about import licenses) and it was late in the afternoon. The sun was low on the horizon—and moving slowly above it from north to south was a line of brilliant silver disks. They looked like metallic mirrors, and they were oscillating or flipflopping with a regular seesaw motion. Once again, I could not guess their size or distance; they were so bright and tiny against the darkening sky that it was also impossible to decide their shape, but they gave the impression of being ellipses. I don't mind admitting that in the few minutes before they came closer I felt myself wondering if the Martian invasion had started; this was the only time I have ever seen a fleet of textbook flying saucers.

In this case, the explanation turned out to be something I already knew—and didn't believe. Many UFO sightings (including one that is the subject of a celebrated and authentic film) were due, I'd read, to birds reflecting sunlight under unusual conditions of illumination. This theory seemed so absurd that I had dismissed it contemptuously; but it is perfectly correct. The lights I saw flipping across Brisbane were nothing more than sea gulls, the undersurfaces of their wings acting as mirrors. Though I have lived beside the sea for a quarter of my life, and am doing so now, this is the only time I have ever witnessed this phenomenon, and I would never have credited it without the evidence of my own eyes. The effect of oscillating metallic disks was absolutely realistic; it would have fooled anyone.

The only UFO that has ever given me the queasy, yet at the same time exhilarating, sensation of being in the presence of the unknown and the inexplicable also occurred in Australia. Perhaps the spectacular surroundings contributed to the impact, for I was standing in Sydney Harbor, immediately beneath the piers of the world's most impressive bridge. (Sorry, San Francisco: size and grace, I'll grant you, but for monumental, built-for-eternity grandeur, nothing can touch Sydney's steel rainbow.)

It was a beautiful, sunny day, and I was looking across the waters of the harbor toward the city, most of which lay framed within that tremendous arch. A strong breeze was sending half a dozen sailing boats scudding over the blue waters and was also driving a few clouds low across the city. But suddenly I realized, with a distinct prickling at the back of the neck, that there was one exception. A single cloud, darker and more compact than its fellows, was floating *completely motionless,* and quite isolated from any buildings, a hundred feet or so above the roof tops.

It was a couple of miles away, and though I stared at it for a good ten minutes it refused to give up its secret. It simply sat in the sky, defying the wind, while all the other clouds went racing past it. There was nothing to do but to hurry back to the apartment and collect a pair of binoculars, hoping that the apparition wouldn't vanish during my absence.

Luckily, it was still there when I returned; through the glasses I could see that it was a hundred feet or so downwind of a tall smokestack, and though there was no visible connection between the two, it was obviously produced by material streaming from this chimney, and condensing as it cooled off. Everyone is familiar with the way in which hot steam leaves the spout of a kettle as an invisible gas and appears a fraction of an inch away in a mist of water droplets. This must have been a similar phenomenon on a slightly larger scale. The gas, vapor, or whatever it was pouring from the chimney condensed for a few seconds as it flowed along the wind, then dispersed again to produce the illusion of an unmoving cloud. In the binoculars it looked rather like a banner flying without benefit of flagpole—or, rather, mysteriously separated from it by a hundred feet of space. Even after I'd worked out the explanation, it was a distinctly uncanny sight.

This strange cloud in the antipodes brings me naturally to another which I once saw much nearer home, above the farm in the west of England where I spent most of my childhood. On this occasion the explanation was immediate and obvious, if you knew the answer—and utterly unguessable if you did not. That many people don't is proved by the fact that one book on flying saucers has made great play of an identical sighting.

Across twenty or thirty years, some of the details are now blurred in my memory, but I am fairly sure that it was early on a bright spring morning, with the dew fresh upon the ground. A gentle wind was blowing, and it was carrying overhead something which I can best describe as an aerial jellyfish. Sometimes it was almost invisible as it turned and twisted in the breeze; at other times the sunlight glanced from its translucent material, so that it looked like a milk-white ghost as it drifted down the sky, being torn apart by the winds even as it moved. I never saw its like again, though it is one of Nature's commoner marvels, familiar enough to those who do not spend their lives locked up in cities.

This silken cloud is something that has baffled mankind for centuries, and even within the last few years it has given rise to the most absurd speculations about the physiology of extraterrestrial visitors. But it is actually the product of a very humble terrestrial creature—the spider. Most young spiders begin their careers as aeronauts, spinning out long threads known as gossamer, which drag them up into the sky on rising air currents. (There is no such thing, incidentally, as a specific gossamer spider; almost all spiders emigrate by air.) On rare occasions, usually in the late summer, the countless threads intertwine to form evanescent clouds, which assume the most extraordinary appearances as the sunlight catches them; when the spiders eventually descend, acres of ground may be covered with their discarded parachutes.

Of all my UFOs, the most beautiful occurred during the war. The time was the summer of 1942, the place a radar station on the east coast of England. It was a perfect, cloudless afternoon— and extremely peaceful, for the blitz was over and the V-weapons had yet to come. If you searched carefully, you could see the pale crescent of the Moon, nearing its first quarter, looking lost and lonely in the daylight sky.

And once you had found the Moon, you could hardly miss what was close beside it—a brilliant, pure white point of light, shining steadily as a star, where no star could be in the sun-drenched sky. Compared with the pallid lunar crescent, it was almost dazzlingly bright, a fraction of a degree away from the Moon, and apparently motionless with respect to it. However, after you had been watching for about ten minutes, you would have noticed that it was moving very slowly toward the Moon—until at last, an hour or so after the first sighting, it finally reached the edge of the chalky lunar disk and merged into it.

The whole sequence of events occupied most of the afternoon, and, as I had an astronomical telescope with me at the station, the conduct of the war was suspended while all the operators and radar mechanics had a close-up view of something that I do not think they will ever forget—and which, if they had seen it for the first time a few years later, they would very likely have interpreted as a flying saucer making a landing on the Moon.

This UFO brings us into the realm of astronomy. When I used the phrase "shining steadily as a star where no star could be" I was technically correct, but deliberately misleading. No stars are bright enough to be seen in the daylight sky, but there is one planet brilliant enough to challenge the Sun. This is Venus, who is easily visible during daytime for the greater part of every year, if you know exactly where to look for her. All down the centuries people ignorant of astronomy have suddenly spotted her in daylight and raised a great hullabaloo, unaware of the fact that they were seeing as commonplace a celestial object as the Moon. (Incidentally, a surprising number of people don't realize that the Moon is visible during the day—still less Venus!)

The sight I observed from the radar station was one of the most striking of astronomical phenomena, and not a particularly rare one. In the course of its movement round the Earth, the Moon is continually getting between us and the other heavenly bodies, partially or wholly hiding them from us. When this occurs with respect to the Sun, we call it a solar eclipse; when the Moon passes in front of a planet or star, it is known as an occultation.

What I have described was an occultation of Venus, seen during the daytime. Though both bodies were moving, most of the apparent motion was due to the Moon in its path around the Earth. About an hour later, Venus emerged from the other side of the Moon and was shining just as brightly as before.

At this point, I would like to pause for a summing-up. Even these few examples collected by one not-very-observant sky-gazer over a period of some twenty years show how extremely easy it is to misinterpret quite ordinary objects when they are seen under unusual conditions. And unless one can arrive at an explanation at the time, there is often no hope of settling the matter at a later date, It remains an unsolved and unsolvable mystery. A perfect example of this was provided a few years ago when an agitated gentleman phoned the police late one night with the news that a flying saucer was racing round his back garden, spitting sparks and flames. When the skeptical cops arrived it was still performing, and after a brief chase they managed to capture it. In a million years, no one—but no one—would guess what it turned out to be. Somebody had been burning trash in a nearby garden, and in the rubbish was an old golf ball. Now a golf ball is highly combustible, and its tightly wound rubber bands contain a great deal of energy—all of which comes out when they start to burn, with the result that the thing takes off like a rocket. Try it one night if you want to scare the neighbors.

Nothing that has so far been said either proves or disproves the existence of genuine, 100 per cent flying saucers from outer space; it merely indicates the need for extreme care in coming to conclusions about peculiar objects seen in the sky. Many UFOs reported by apparently reliable observers are quite inexplicable in terms of current knowledge, but even this does not prove that they are necessarily the products of intelligence—terrestrial or otherwise. For there is now no doubt that when Nature *really* tries, she can produce "spaceships" that would satisfy the most exacting requirements.

Here is the proof: I am quoting from the May 1916 issue of *The Observatory,* a scientific journal published by the world's leading astronomical organization, the Royal Astronomical Society. The date—1916—is important, but the description is of an event that occurred more than thirty years before, on the night of November 17, 1882.

The writer was a well-known British astronomer, Walter Maunder, then on the staff of the Greenwich Observatory. He had been asked to describe the most remarkable sight he had ever seen during his many years of observing the heavens, and he recalled that soon after sunset on that November night in 1882 he had been on the roof of the observatory, looking across London when:

A great circular disc of greenish light suddenly appeared low down in the East-North-East, as though it had just risen, and moved across the sky, as smoothly and steadily as the Sun, Moon, stars and planets move, but nearly a thousand times as quickly. The circularity of its shape was merely the effect of foreshortening, for as it moved it lengthened out, and when it crossed the meridian and passed just above the Moon its form was almost that of a very elongated ellipse, and various observers spoke of it as "cigar-shaped," "like a torpedo" . . . *had the incident occurred a third of a century later, beyond doubt everyone would have selected the same simile—it would have been "just like a Zeppelin."* [My italics.]

Remember that Maunder was writing this in 1916, when Zeppelins were very much in the news—even more so than spaceships are today.

Since hundreds of observers all over England and Europe witnessed this object, reasonably accurate figures for its height, size, and speed were obtained. It was 133 miles above the Earth, moving at 10 miles a second—and must have been at least 50 miles in length.

What was it? No one could have given a full answer to that question in 1882, but today we can do so with complete confidence. The solution follows from a clue which I have deliberately omitted; the object was seen during a violent auroral display and was undoubtedly part of it.

We now know that auroras are caused by streams of electrified particles shot out of the Sun which cross space and eventually enter the Earth's atmosphere. Here they produce a type of fluorescence much like that which lights up our neon tubes and gas-discharge lamps. Billions of years before Broadway, Nature was hanging her illuminated signs in the polar skies.

Though the Sun is the original source of the energy, our planet is responsible for the strange shapes that the aurora assumes —its ever-changing streamers, curtains, and rays. For the Earth's very weak but far-ranging magnetic field, extending thousands of miles out into space, has a focusing effect on these streams of particles, concentrating them at the poles. It makes them paint pictures on the sky, as very similar beams and magnetic fields produce images on the screens of our TV sets.

And sometimes, surprising though it seems, Nature with her 93,000,000-mile-long TV tube can create apparently symmetrical, sharp-edged objects moving steadily across the heavens. (Maunder specifically states that the phenomenon he observed "appeared to be a definite body.") This seems much more remarkable to me than any mere spaceship, but the facts are beyond dispute. Observations of the "torpedo" through the spectroscope proved its auroral nature, and as it passed across Europe it slowly began to break up. The cosmic TV tube went out of focus.

It may be argued that this weird—possibly unique—event cannot account for the hard core of unexplained UFOs, many of which have been observed in the daytime, when the faint light of the aurora is invisible. Yet I have a hunch that there is a remote connection, and this hunch is based upon a new science which has developed during the last few years, largely under the impetus of missile and nuclear research.

This science—take a deep breath—is magnetohydrodynamics. You'll be hearing a lot more of it in the future, for it's one of the keys to space exploration as well as atomic power. But it concerns us here only because it deals with the movement of electrified gases in magnetic fields—with the sort of thing, in fact, which startled Mr. Maunder and a few thousand other people in 1882.

Today we call these objects "plasmoids." (A lovely word, that. Can't you see the title from some he-man's magazine of the Space Age: "I Was Pursued by Plutonian Plasmoids"?) They've been known for quite a while, in the form of one of the most baffling phenomena in the whole of Nature—ball lightning, which is something no one would believe without overwhelming evidence. During thunderstorms, brilliantly glowing spheres are sometimes seen rolling along the ground or moving slowly through the air. Occasionally they explode with great violence, and so until recently have all the theories put forward to explain them. But now we have been able to make small versions—baby plasmoids—in the laboratory, and there have been horrid rumors that the military are trying to develop them as weapons.

I have never seen ball lightning and am by no means sure that I want to, at least at close quarters. However, with this example of the fantastic tricks natural forces can play, it would be very unwise to argue that even the most impressive UFO *must* be artificial. In fact, a good working rule for UFO observers is: It's not a spaceship unless you can read the Mars registration plate.

Of course, some people claim to have done a good deal better than this, but luckily I am not concerned here with the more extreme aberrations of the human mind. The saucer mania of our age will provide a fascinating study for future psychologists; I find it not amusing but saddening. I could hardly raise a smile when a good lady in Pennsylvania recently attacked me for disbelieving in flying saucers, giving as evidence the fact that they were continually landing in her garden. They made, she added, quite a lot of noise—though the only sound she had definitely identified was "a beautiful, long-drawn-out hallelujah. . . ."

Since one can never rule out *all* possibilities, there must always remain the faint chance that some UFOs are visitors from elsewhere, though the evidence against this is so overwhelming that it would require an article much longer than this to give it in detail. And if this verdict disappoints you, I can offer what seems to me very adequate compensation.

If you keep looking at the sky, before much longer you *will* see a genuine spaceship.

But it will be one of ours.

Postscript

Since writing the above, I have seen the finest—and most "classical"—flying saucer of my life. On October 17, 1958, I was aboard KLM Flight 826, coming up the coast of Italy on a bright but somewhat hazy afternoon. We were at about ten thousand feet, en route to Geneva, and the ground was clearly visible at the time (around 2 P.M.).

I was looking down at the coastline almost immediately beneath us, waiting for Naples and Vesuvius to come into view, when I became aware that a brilliant oval of light was keeping pace with the aircraft a few thousand feet below. It appeared to be quite solid, though its edges were hazy and seemed to pulsate slightly;

they also had a bluish tinge rather like that of a mercury arc. It was impossible to judge its size or distance, but I had the impression that the object was halfway between the aircraft and the ground. Sometimes it was so brilliant that it hurt the eye to look at it directly.

It was in view for a good ten minutes, keeping station beneath us, and for long periods of time it was remarkably constant both in shape and size. Apart from the occasional quivering of its edge, there was no way of telling that it was not a solid disk; it completely masked the ground beneath. Several of my fellow passengers were busy photographing it, and I am quite sure that they are now proudly showing *genuine* photos of flying saucers to their friends.

I must confess that had I caught only a glimpse of this apparition I should have been quite baffled; as it was, I was able to keep it in sight until it disintegrated and slowly faded from view, like a cloud breaking up beneath the Sun. By that time there was no question of its identity.

It was a mock sun, or "sun dog," caused by the presence of an invisible layer of ice crystals between the aircraft and the ground. They are fairly common, though this is the first I have ever seen. The ice crystals act as tiny mirrors, each reflecting an image of the Sun; the combination of myriads forms the brilliant disk which, being a reflection, appeared to follow the aircraft. D. H. Menzel's book *Flying Saucers* has a fine photograph of a mock sun as its frontispiece; the one I observed was more sharp-edged and must have been formed in an unusually stable layer of air, in which the vast majority of ice crystals had almost the same orientation.

Postpostscript

Little did I dream when I wrote the preceding article, in 1958, that UFOs would still be flourishing, though perhaps not as energetically, more than a decade later. . . . In the intervening years perhaps the most important development has been the official

United States Air Force study, which culminated in the bitterly disputed "Condon report." The conclusions of that report—which, predictably, were not accepted by UFO believers—were that though a few sightings were still unexplained (and quite mysterious) they did not merit a continued large-scale investigation. A small handful of qualified scientists disagree with the Condon report and regard the "extraterrestrial" hypothesis as the *least* improbable explanation of the more baffling cases.

And meanwhile, I have seen my most convincing UFO: see "Son of Dr. Strangelove," page 238.

V. *Son of Dr. Strangelove, Etc....*

Which Way Is Up?

20

When I became amphibious, I never expected that it would cause such confusion among my friends. Yet I can understand their feelings; when one has been writing and talking about space flight for the best part of twenty years, a sudden switch of interest from the other side of the stratosphere to the depths of the sea does seem peculiar. It might be regarded as a serious failure to keep to the point—even a demonstration of a certain lack of stability. So, to put the record straight, I'd like to explain just why it is that I've traded in my space suit for an aqualung, my telescope for an underwater camera.

The first excuse I give to baffled journalists and lecture chairmen agonizing over their introductions is the economic one: submarine exploration is so much cheaper than space flight. The first round-trip ticket to the Moon is going to cost at least $10 billion if you include research and development. By the end of this century it will be down to a few million, but the complete basic kit needed for skin diving (flippers, face mask, and snorkel tube) can be bought for $20. Which, it can hardly be denied, is a very modest price to pay for admission to a new element.

My second argument is more philosophical: the ocean, surprisingly enough, has many points of similarity to space. Some of these I had guessed even before I first went underwater; others

219

I did not discover until I had been diving for several years, though I do my best to claim that I had anticipated them all.

In their different ways, both sea and space are equally hostile to man. If we wish to survive in either for any length of time, we have to employ mechanical aids. The diving dress was the prototype of the space suit; the sensations and emotions of a man beneath the sea will have much in common with those of a man beyond the atmosphere.

One of those sensations is weightlessness, and it was this fact, as much as any other, that first got me interested in underwater swimming. Here on the surface of the Earth, it is never possible to escape from gravity. All our lives, we creatures of the land must drag the weight of our bodies around with us, envying the freedom of the birds and clouds.

In a spaceship, however, once the thrust of the rockets has ceased, all weight vanishes, and the effect that this will have upon the human organism has long been the subject of debate among medical men. It has been suggested that "spacesickness" and perhaps total incapacity might result when there is no longer any way of distinguishing between up and down, because both conceptions have lost all meaning.

Something rather like this happens underwater, for gravity plays little part in the lives of fish and other marine creatures. Looking at the matter scientifically, it occurred to me that, if I imitated them, I might discover what it felt like to be a spaceman.

There is no doubt that one of the greatest attractions of skin diving is the sense of freedom in three dimensions that it gives; when your buoyancy is properly neutralized by lead weights, you can float without any effort at any level. If you push against a rock or kick off from the sea bed, you will drift slowly until the friction of the water destroys your momentum. Until the first manned satellite is established, this is the nearest we shall come to knowing the conditions that prevail inside a spaceship.

I soon discovered, however, that the analogy was not exact. Though you possess no weight when you are submerged, up and down still exist. Even when the other senses fail, your eyes can give you all the orientation you need. Unless you are swimming after sunset, or in very dirty water, you can always tell the direction from which the light is coming. It may be no more than a vague glow, like the first hint of dawn, but it is an unmistakable signpost to the surface.

Well—almost unmistakable, for there are exceptions even to this rule. I was once diving in a somewhat gloomy coral cave whose floor was covered with light sand when I was surprised to see that most of the fish around me were swimming upside down. All the light came from below, and they'd been fooled into thinking that this direction was up.

Men are, on the whole, more intelligent than fish, but here is a case where what counts is instinct, not intelligence. It would seem that as long as the cabin of a space vehicle looked normally oriented to the eye, the danger of vertigo would be greatly reduced, even in the complete absence of gravity. However, if chairs and tables were bolted indiscriminately to all six walls, that would be asking for trouble. Even the most hardened astronaut might soon feel unhappy unless there was a general agreement that a certain direction would be up, and the cabin was designed and used accordingly. (One can picture the warning notices PLEASE DO NOT SIT ON THE CEILING.) Once the eye was satisfied, its signals would override any messages from other sense organs which were frantically telling the brain that gravity had ceased to exist.

It was Cousteau who coined the phrase "Silent World" to describe the sea, but the description is even more applicable to space. There are a few sounds underwater: porpoises squeak, whales groan, shrimps snap their claws. In the vacuum of space, however, no sounds can exist, for there is nothing to transmit them. The only noises that a space traveler will normally hear are those created inside his ship—the whirring of electric motors, the hiss of air pumps, the clank of metal upon metal. These sounds would echo around and around the little world of the ship and would form a continuous background which would be noticed only when there was some change in it. In the same way, an aqualung-er is seldom consciously aware of the bubbling of his exhaust valve, but when it stops he reacts at once, even before he feels the alteration in the air flow.

Very occasionally, a space traveler might hear a noise from the outer world. From time to time particles of meteor dust would hit the hull with enough impact to make an audible sound; on still rarer occasions, when the meteor was a really large one, that sound might be the last thing that the voyager would ever hear.

In space there are no horizons; the questing eye reaches out forever, in all directions, and finds no fixed point at which to rest. For this reason there is also no real sense of distance; in the absence of perspective, it is impossible to judge the remoteness of the stars. They could be pinpoints of light a few miles away, as indeed the ancients thought they were. The truth is so incredible that the instinct rejects it, and a man midway between the planets might feel that he could reach out and grasp the gleaming sparks around him.

This sense of floating in a void which is not infinite, but merely indefinite, is one that can be captured in the sea under certain conditions. If you dive into deep water and head quickly downward, you can lose all sight of the surface before there is any trace of the bottom. You will then be suspended in a completely featureless blue-green void, and if there are no fish within your field of vision it is quite impossible to judge how far you can see. Visibility may be a hundred feet, yet you can delude yourself into thinking that you cannot see more than a yard.

This is not a pleasant sensation, and more than once I have been glad to reassure myself, simply by stretching out my hand and looking at my fingers, that I *could* see farther than my nose. Whether a similar illusion will arise in space we will not know until we can get a few million miles away from Earth; if it does, then the ocean is one place where we can prepare men to meet it.

Another lesson for space which I have learned from the sea is that the human body is much tougher and more adaptable than anyone could reasonably have expected. Although it is necessary, in a vehicle traveling beyond the atmosphere, to provide complete protection against the vacuum of space by the use of a pressure cabin, I believe that the achievements of today's skin divers have demonstrated that men could withstand even exposure to airless space for appreciable periods of time—a fact which might make all the difference between life and death in any emergency.

This statement will certainly astonish a good many people, especially those who have read science-fiction stories containing gruesome accounts of what happens to space travelers when their ship springs a leak or is punctured by a meteor. Yet in either of these cases, it would normally take several seconds for the air pressure to drop to zero, and a skin diver coming up quickly from a depth of only ten feet experiences a greater pressure drop, in a far shorter time, than the occupants of a spaceship would undergo if their vessel was suddenly punctured.

Skin diving has also shown what an extraordinarily long period of time men can exist without breathing, if they have suitable training and preparation. The first time I went underwater, I stayed down all of ten seconds. But as I became more confident and learned the tricks of the trade, I was able to push my endurance up to three and a half minutes; though this sounds impressive, it is nothing compared with the record, which is now over thirteen minutes.

This has convinced me that trained men, given sufficient warning so that they could prepare themselves, would be able to stand exposures of a minute or so even to the vacuum of space. Recently I had a chance of arguing this point with Major David Simons, the only man who has so far spent more than a day beyond the effective limits of the atmosphere. (During his famous balloon ascent in 1957, he had more than 99 per cent of the atmosphere below him, so that for most physiological purposes he was out in space.) Major Simons was willing to grant me that a man could remain conscious for fifteen seconds on exposure to vacuum, but he thought that death would then follow swiftly because the brain would be deprived of oxygen.

Well, fifteen seconds is a very long time in an emergency—long enough to get into the next cabin and slam the airtight doors. And I have a hunch that the margin of safety may be better than fifteen seconds, for the human body has so often surprised us in the past by its unexpected powers of adaptation. Not long ago, doctors proved conclusively that a naked diver could not possibly descend 100 feet without having his lungs crushed by the pressure. Yet the skin-diving record is now 140 feet *without breathing gear,* and there is evidence that some divers have been down 200 feet—a depth at which the pressure on every square foot of the body is over five tons. Yes, the human frame can take a lot of punishment if it has to, and there are occasions when a space pilot may be tougher than his ship.

In the exploration of a new element, psychology is as important as physiology. From my own experience, I'm convinced that under-water exploring inculcates the kind of mental outlook which we shall need in space. It may be summed up as a sense of alertness— a realization that almost anything can happen, and that when it does you've got to be ready for it. This is not a question of being nervous or apprehensive, so much as being prepared, so that you react properly and don't panic. In the sea, panic can be the deadliest of killers, and it needs so little to bring it on—a strange movement glimpsed out of the corner of the eye, a slight malfunctioning of equipment, a shadow crossing the sea bed when you know there are no clouds in the sky, a sound in a world which is normally silent. And, above all, an unexpected, purposeful contact when you think you are floating alone in mid-ocean. . . .

There is a test that the Australian Navy used on its frogmen to separate not the men from the boys but the men from the super-men. (Readers prone to nightmares had better skip the next two paragraphs.) It consisted of sending a trainee down into the water, at night, with his face mask blacked out so that he was totally blind. A second diver with a sealed-beam searchlight would be in the neighborhood to keep an eye on the victim, who had been instructed to swim back to the surface. This is not difficult, even when you cannot see your way, because it is a simple matter to increase buoyancy and thus go up like a balloon. In this case, however, there was a fiendish complication of which the victim was unaware.

He had been released in the middle of an underwater jungle, a dense forest of kelp. The thin fronds, scores of yards long, formed a close-packed wall around him, and the current carried him steadily toward it. Without the slightest warning he would hit this floating barrier, and at once the tons of unstable vegetation would collapse, engulfing him (in utter darkness, remember) beneath an animated avalanche of twining tendrils. By the time he had been dug out of this and brought back to the surface, his instructors would know if he'd made the grade.

Anyone who could pass a test like this would be a useful man to have around in one of those typical space emergencies where the atomic pile is about to go out of control, the captain is down with the D.T.s, the last of the oxygen is leaking through a meteor puncture, and the Thing has broken loose from its cage in the hold.

Talking of Things leads us to another, and somewhat speculative, link between sea and space. Sooner or later, during our exploration of the Universe, we are going to encounter utterly alien forms of life. It does not seem likely that we will meet them on the Moon when we get there in the 1970s, but the first contact may occur on Mars a decade or so later.

There is absolutely no way of guessing what shape extraterrestrial life forms may take; even if we had perfect knowledge of conditions on Mars and Venus (the only planets where protoplasmic life could exist), we would be no nearer to picturing the creatures that might live there. If anyone doubts this, let him ask himself if he could have predicted the elephant, the duck-billed platypus, the giraffe, or *Homo sapiens* from a geophysical survey of the planet Earth.

Until we reach them—or they reach us—we shall remain in

complete ignorance about the creatures which may exist on other planets. Perhaps we may find no more than a few lichens on Mars; perhaps our first encounter with extraterrestrial animals or intelligences may still lie centuries in the future. Yet even now, by sinking down into the sea, we can capture many of the sensations our descendants will know when they set foot upon other planets. Certainly nothing that they will ever meet there can be more fantastic than some of the creatures that inhabit the waters of this world.

This is another reason why underwater exploring is, psychologically, a good preparation for man's adventure in space—and why, incidentally, it can be a good corrective to the psychotic horror movies which depict all extraterrestrial beings as hideous monsters bent on destruction. Monsters do not exist in Nature, but only in men's minds. I learned this lesson the first time I met a giant manta ray, and I have never forgotten it.

Sometimes known as the devilfish, because of its grotesque batlike shape and the two horns, or palps, extending on either side of its mouth, the manta is one of the weirdest looking beasts in the sea. When, long before I had dreamed of doing any underwater exploring myself, I saw some of Hans Hass's photos of this strange creature (which can grow up to thirty feet across) I thought I had never seen anything so hideous; its head reminded me strongly of the gargoyles on Notre Dame.

Yet, five years later, when I encountered one of the great beasts peacefully browsing over a coral reef off the Queensland coast, that initial feeling of repulsion vanished completely. Here, it was true, was something strange and beyond ordinary experience, but it was no longer hideous—it was not even alien. Its fitness of purpose and the grace of its movements as it flapped along the reef, keeping a wary eye on the human invaders of its territory, left little room in my mind for anything except admiration—and a furious rage against those fishermen (above or below the water) who sometimes spear these huge, harmless beasts for their amusement.

To most people, perhaps the most ghastly inhabitant of the

sea—the ultimate in creeping, malevolent horror—is the octopus. The very thought of contact with its slimy, sucker-studded tentacles is enough to make them feel physically sick, yet once again this is a reaction founded on ignorance or inspired by stories put out by divers who want to make their job sound even more dangerous than it is. I would not go so far as to say that the octopus is a friendly, attractive beast which no home should be without, but I would claim that almost all one's original revulsion vanishes when one gets to know this talented mollusk. In real life, though not when seen frozen in menace by an imaginative illustrator, the octopus is quite fascinating to watch as it jets across the sea bed or slithers briskly from rock to rock, only too anxious to keep out of your way. And its rapid color changes, when it is excited or nervous, are really beautiful.

These examples should be sufficient to prove my point—that there is nothing in the natural world, however strange it may be, that one cannot grow accustomed to. Albert Schweitzer must have had this in mind when he formulated his doctrine of "reverence for life"; it is a creed that a man of sensitivity can learn in the sea as nowhere else, and it is one which mankind must master before it makes contact with other intelligent races in the Universe. I have never been convinced that intelligence comes only in one model—and that that model has two legs, two eyes, and one mouth.

Someday we may encounter representatives of far higher civilizations than ours, who may differ from us as greatly as we differ from the manta or the octopus. And as we have to overcome color prejudice, so our descendants may have to overcome a much more fundamental *shape* prejudice. The time may come when no well-bred person would dream of remarking that the ambassador from Rigel looks like a cross between a jellyfish and a tarantula (even if he does) or would be particularly upset because the members of the Sirian trade delegation have not only three heads but also four sexes.

Fantasy? Of course; the reality of our Universe *is* fantastic. We live in an age when we can keep up with tomorrow—or even today—only by letting our imaginations freewheel anywhere they care to travel, as long as they keep within the bounds of logic and the known laws of Nature.

Yet, if we hope to reach the stars, we shall need more than imagination, more than scientific skill. These alone would be useless without the spirit of adventure which conquered our own world in the days when much of this Earth was as mysterious and remote as the planets seem today.

That spirit is not lacking; along all the coasts of the world, boys (and girls) barely in their teens are setting off on subaqueous journeys which would have seemed utterly incredible to their grandparents, and which must often terrify their parents. Among those youthful skin divers, the men who will make up the space crews of tomorrow are already learning courage, judgment, self-confidence, and those less easily defined qualities needed by all great explorers.

I began this apologia on a personal note; I would like to end it on one. The parallels between sea and space are sufficiently clear, and there is no need to say any more to prove that underwater exploration has a perfectly logical tie-in with astronautics. Yet logic is never enough; it was Bertrand Russell—somewhat surprisingly—who remarked that the purpose of reason is to give us excuses for doing the things we want to do.

In the final analysis, I went undersea because I liked it there, because it opened up to me a new, strange world as fantastic and magical as the one that Alice discovered behind the looking-glass. And perhaps I did it because, after hearing people call me a space-travel expert for twenty years, I felt I was getting into a rut. As Hollywood stars know very well, it is fatal to become typed; if you want to progress, to continue your mental and emotional growth, every so often you must surprise yourself (and your friends) by changing the pattern of your life and interests.

Once you are neatly classified and pigeonholed, incapable of any further development, your life is over. You might as well be a stuffed specimen in a museum, completely described by the label tied to your ankle. When there's nothing more to say about you, you're already dead.

I feel very happy to have avoided that fate, but there's one problem that sometimes worries me. What new track do I switch to in 1975?

Postscript

The preceding paper was written in 1957, the first year of the Space Age; my guess of $10 billion for the first round-trip ticket to the Moon turned out to be surprisingly accurate.

Today, the connections between sea and space are widely recognized, and scuba diving is part of the training of all astronauts. In March 1970, with the cooperation of the Ceylon Navy, my partner Hector Ekanayake and I had the great pleasure of taking the Apollo 12 astronauts—Conrad, Bean, and Gordon—diving in the magnificent harbor of Trincomalee, on the east coast of Ceylon.

The free diving record (by a man without scuba gear) now stands at 240 feet. The record for survival in vacuum (by dogs and chimpanzees) is about four minutes, and at least one human being has survived an (accidental) exposure to a vacuum without ill effects.

Haldane and Space

21

This essay was written at the request of Dr. K. R. Dronamraju, for the excellent memorial volume he edited—*Haldane and Modern Biology* (Johns Hopkins Press, 1968). Haldane has also been the subject of a fine biography by Ronald Clark: *J. B. S.: The Life and Work of J. B. S. Haldane* (Hodder & Stoughton, 1968).

Carl Sagan's speculations on direct contact between stellar civilizations may be found in the stimulating book he wrote with Iosef Shklovskii, *Intelligent Life in the Universe* (Holden-Day, 1966). By no coincidence whatsoever, this book is dedicated "To the memory of John Burdon Sanderson Haldane, F.R.S., member of the National Academies of Science of the United States and of the Soviet Union, member of the Order of the Dolphins, and a local example of what this book is about."

Professor J. B. S. Haldane was perhaps the most brilliant scientific popularizer of his generation; starting in 1924 with *Daedalus, or Science and the Future,* he must have delighted and instructed millions of readers. And, unlike his equally famous contemporaries Jeans and Eddington, he covered a vast range of subjects. Biology, astronomy, physiology, military affairs, mathematics, theology, philosophy, literature, politics—he tackled them

230

all. He also wrote a workmanlike novella, *The Gold Makers,* and a charming tale for children, *My Friend Mr. Leakey.*

Although some are naturally dated by the progress of science, most of Haldane's scores of essays (they appeared in such varied places as *Harper's Magazine, The Saturday Evening Post, The Strand Magazine, The Spectator, The Daily Express,* the St. Louis *Post-Dispatch*—and, of course, *The Daily Worker*) may still be read with great profit. Some of the volumes in which they were reprinted, such as *Possible Worlds* (1927) or *The Inequality of Man* (1932) may now be difficult to obtain. However, one of Haldane's most famous essays, "On Being the Right Size," is readily available in Volume 2 of James Newman's *The World of Mathematics.* It is a perfect example of his lucidity and the breadth of his interest.

As far as I can recall, I was first attracted to Haldane's writings by the element of extrapolation they contained. He was obviously sympathetic to science fiction and astronautics; indeed, I have just discovered this paragraph in his very first book, *Daedalus*:

I should have liked had time allowed to have added my quota to the speculations which have been made with regard to inter-planetary communication. Whether this is possible I can form no conjecture; that it will be attempted I have no doubt whatever.

It was through space flight that I made my first, and somewhat alarming, encounter with Professor Haldane. In 1951, as Chairman of the British Interplanetary Society, I invited him to give our organization a paper on the biological aspects of space flight. Despite the very short notice (the lecture was a substitute for one by Professor J. D. Bernal, which had to be postponed to a later date), Haldane at once agreed to step into the breach.

He and Helen Spurway (later Mrs. Haldane) duly arrived at the Caxton Hall, Westminster, in one of the most decrepit cars I have ever seen; it appeared to be held together largely by rust. When I greeted him at the top of the steps and reached out to take his hat, he insisted on retaining it for sanitary reasons. The cat, he explained, had just used it for an unauthorized purpose—though he put the matter more pithily, if you will excuse the pun.

After this not very auspicious beginning, the lecture was a great success.* It dealt with three problems: how men would live in spaceships, how they would live on other planets, and what sort of life they might find there. In 1951, these were not subjects with which many reputable scientists cared to be associated, and Haldane himself had sometimes adopted a rather conservative attitude toward space flight. In his remarkable essay "The Last Judgement" † he set the first landing on Mars in the year 9,723,-841, and an expedition to Venus "half a million years later." This demonstrates once again how hard it is for even the most farsighted scientist to anticipate the future. Haldane could hardly have believed, in 1927, that he would live to see the Apollo project, and to participate himself in state-sponsored conferences on exobiology.

Though it has naturally been superseded in many respects, Haldane's 1951 paper still contains some interesting ideas. He must have been one of the first to point out the dangers of solar flares and to suggest that space voyages should be made during periods of minimum solar activity. And, with his tongue firmly in his cheek, he suggested that we should take seriously the hypothesis that life has a supernatural origin—from which he concluded that, as there are 400,000 species of beetles on this planet, but only 8,000 species of animal, "the Creator, if he exists, has a special preference for beetles, and so we might be more likely to meet them than any other type of animal on a planet that would support life."

After the lecture, we took Haldane and Miss Spurway to dinner at the Arts Theatre Club, and of the sparkling conversation that doubtless ensued I remember only one item. This, however, involves such a striking and melancholy coincidence that it is worth recording.

* A. E. Slater, "Biological Problems of Space Flight." A report of Professor Haldane's lecture to the Society on April 7, 1951. *Journal of the British Interplanetary Society* X, 4 (July 1951), 154–158.
† In *Possible Worlds* (London: Chatto & Windus, 1927).

Presumably we had been discussing problems of respiration, for Haldane expressed the belief that, in the right circumstances, animals could "breathe" water. One of his reasons for believing this was the observation that it was extremely difficult to drown newborn mice; it seemed that their lungs were still capable of extracting oxygen from water. Haldane then made the sadly prophetic statement: "If I knew that I was dying of cancer, I'd like to make this experiment. It would probably be rather painful. . . ."

Now, Johannes Klystra has demonstrated water-breathing, with animals up to the size of dogs. But Haldane had been thinking of this as far back as 1951.

Our paths did not cross again for almost ten years, when we had both migrated to the East. In November 1960, the Ceylon Association for the Advancement of Science invited Haldane to Colombo to address its annual meeting, and it was characteristic of him that on arrival he immediately abandoned the official hotel in favor of a modest Indian (and vegetarian) hostel in a less fashionable suburb of the city.

I debated for a considerable time before calling on him. In the intervening years, I had heard rumors of his ferocity—some reports of his behavior to journalists made him sound rather like Conan Doyle's Professor Challenger—and I had no idea whether he remembered our last encounter, still less whether I was *persona grata*. Nevertheless, quaking slightly and with my partner Mike Wilson to give me moral and (if necessary) physical support, I called at his hotel and sent up my card.

His first words when he arrived on the scene, dressed in his white gown and looking like a Hindu patriarch, were not very reassuring. "Oh my God!" was distinctly off-putting, and a real or feigned deafness made any further communication appear hopeless. I was about to leave with as little fuss as possible when I suddenly realized that, far from being exasperated at the intrusion, he was genuinely glad to see me. I was not so surprised to discover that he had read most of my books; for Haldane, of course, had read *everything*.

Within a couple of hours the Haldanes had arrived at my house, where the Professor leaped upon my technical library like a starving man. Later in the afternoon we took him on a tour of Colombo's excellent zoo, not knowing that he was suffering from a spinal injury that must have caused him great discomfort. When this was subsequently discovered, he apologized for any absent-mindedness, adding that a fractured vertebra was not all that important as he had "learned to ignore certain types of sensory input."

A few days later, the Wilsons and I invited the Haldanes and their Indian colleagues (Drs. Davies and Dronamraju) to dinner at our house. After this length of time, I can remember only two fragments of small talk. At one point, the Haldanes were demolishing reputations with such gusto that I felt constrained to remark "That's what I like about science—the way it rises above personalities." And when the conversation turned, via Unidentified Flying Objects, to atmospheric electricity, I asked the Professor, "Is it true that, when he had a research station on Pike's Peak, your father did some work on ball lightning?"

He answered at once, "No—ball lightning did some work on *him.*"

After the dinner, we screened Mike Wilson's underwater movie, *Beneath the Seas of Ceylon,* showing the behavior of the teeming population of the Great Basses Reef and in particular recording the intelligence of a family of black groupers (*Epinephelus fuscoguttatus*).* The spectacle of these giant fish cooperating as movie extras impressed Haldane so much that he frequently gave vent to a surprisingly schoolboyish "Golly!"—a term which, for all its naïveté, expresses the sense of wonder that is the hallmark of the great scientist.

We never met again, but our real acquaintance had started, and

* Arthur C. Clarke and Mike Wilson, *Indian Ocean Adventure* (New York: Harper & Row, 1961).

it continued in subsequent correspondence. In April 1962 I received a pressing invitation to stay with the Haldanes, opening with a rather ambiguous compliment: "Allow me to congratulate you on the Kalinga Prize. Personally I should also like to see you awarded a prize for theology, as you are one of the very few living persons who has written anything original about God. You have in fact written several mutually incompatible things. . . . if you had stuck to one theological hypothesis you might be a serious public danger."

To my lasting regret, I was unable to accept Haldane's hospitality, because I had contrived to get myself almost completely paralyzed, and it was some months before I was able to walk again. And when I finally got to the Kalinga ceremony, it was in New Delhi, not Orissa; so I was farther away from the Professor than if I had stayed in Ceylon.

Thereafter, my slow convalescence and a series of other problems prevented a rendezvous, but we continued to correspond hopefully. Haldane's letters, usually handwritten, often ran to a thousand words and were so full of ideas as his agile mind jumped from one subject to another that they were both good fun and hard reading. It was obvious that he took great pride in the team he had built up around him at Bhubaneswar; as he put it: "I seem to have taken the lid off some young men, and they are making really fantastic discoveries."

A few extracts will serve to give the flavor of that final correspondence, covering the period April 12, 1962, to January 8, 1964.

I want to talk to you seriously about the soul and all that.

You have been listening to the apiary in Professor J.B.S. Haldane's bonnet.

It is clear that a gibbon, and still more a prehensile-tailed South American monkey (or a man-sized version of one) is much better pre-adapted than *H.sap* for low gravitational fields . . . we might get back these lost appendages by intranuclear grafting. We should then find it natural to count up to 210 (10 fingers) $\times$ (10 fingers $+$ 10 toes $+$ 1 tail). This would be a better base than 10 (being $2 \times 3 \times$

5 $\times$ 7) and a slight improvement either in cerebral organisation or teaching methods would enable people to learn the necessary multiplication table.

I suspect the hymenoptera and isoptera are the best hope for studying non-human technology. My wife, for reasons of her own, regards the diptera as Top animals.

I have been thinking about cosmonautics (i.e. going to Alpha Centauri and further). It seems to me that there are two possibilities. (1) It is practicable to reach speeds of the order of ½ that of light. (2) To avoid too highly energetic collisions with dust clouds, it is not practicable to exceed about 1,000 km/sec. which is near the upper bound of relative stellar velocities in our neighborhood. As there are probably a lot of animal species in the galaxy with a technology more advanced than ours, but they don't seem to visit our planet often, I suggest that (2) is the more probable. If (1) is more correct, the voyages would mainly be undertaken at speeds near that of light. . . .

An intelligent species is pre-adapted for interstellar travel if (a) it is very long-lived or reproduces clonally, so that the crew will have the same set of personalities after the numerous generations needed to travel long distances, and (b) it is accustomed to a very large gravitational field. If white dwarfs cool down, and life evolves on them, their inhabitants, though nearly two-dimensional, could be boosted with an acceleration which would flatten you and me. Has this point been made? If not, it is a present for you.

The very last letter I ever received from Haldane was written from University College Hospital on January 8, 1964, and characteristically it linked his final illness with the subject of astronautics. After describing his present plight after his colostomy he remarked "I (and a million other surgical cases) would be quite satisfied with lunar surface gravitation (1/6g). A few would no doubt be better in free fall. . . ."

The same letter reverted to our earlier discussion of interstellar flight. During his visit to the United States, Haldane met Carl Sagan, who had given him his stimulating paper on direct contact between galactic civilizations.* This obviously inspired these spec-

* Carl Sagan, "Direct Contact Among Galactic Civilizations by Relativistic Interstellar Spaceflight," *Planetary and Space Science* XI (1963), 485–498.

ulations: "I suggest the following hypotheses. Interstellar travel occurs on a vast scale. 'Cosmic rays' are merely the exhaust of rockets. The rocketeers do not often visit us for any of several reasons. They may think we are nasty chaps. They may mostly be social arthopods, uncertain how to help members of a different phylum to evolve or behave. And so on."

The length, cheerfulness, and intellectual energy of this letter completely deceived me. Haldane had always seemed indestructible, and I continued to plan for our meeting in Orissa.

It was a great shock to hear of his death a few months later, and to realize that communication had at last broken with the finest intellect it has ever been my privilege to know.

Son of Dr. Strangelove

Or, How I Learned to Stop Worrying and Love Stanley Kubrick

22

The first steps on the rather long road to *2001: A Space Odyssey* were taken in March 1964, when Stanley Kubrick wrote to me in Ceylon, saying that he wanted to do the proverbial "really good" science-fiction movie. His main interests, he explained, lay in these broad areas: "(1) The reasons for believing in the existence of intelligent extra-terrestrial life. (2) The impact (and perhaps even lack of impact in some quarters) such discovery would have on Earth in the near future."

As this subject had been my main preoccupation (apart from time out for World War II and the Great Barrier Reef) for the previous thirty years, this letter naturally aroused my interest. The only movie of Kubrick's I had then seen was *Lolita,* which I had greatly enjoyed, but rumors of *Dr. Strangelove* had been reaching me in increasing numbers. Here, obviously, was a director of unusual quality, who wasn't afraid of tackling far-out subjects. It would certainly be worthwhile having a talk with him; however, I refused to let myself get too excited, knowing from earlier experience that the mortality rate of movie projects is about 99 per cent.

Meanwhile, I examined my published fiction for film-worthy ideas and very quickly settled on a short story called "The Sentinel," written over the 1948 Christmas holiday for a BBC contest. (It didn't place.) This story developed a concept that has since

238

been taken quite seriously by the scientists concerned with the problem of extraterrestrials, or ETs for short.

During the last decade, there has been a quiet revolution in scientific thinking about ETs; the view now is that planets are at least as common as stars—of which there are some 100 billion in our local Milky Way galaxy alone. Moreover, it is believed that life will arise automatically and inevitably where conditions are favorable; so there may be civilizations all around us which achieved space travel before the human race existed, and then passed on to heights which we cannot remotely comprehend. . . .

But if so, why haven't they visited us? In "The Sentinel," I proposed one answer (which I now more than half believe myself). We may indeed have had visitors in the past—perhaps millions of years ago, when the great reptiles ruled the Earth. As they surveyed the terrestrial scene, the strangers would guess that one day intelligence could arise on this planet; so they might leave behind them a robot monitor, to watch and to report. But they would not leave their sentinel on Earth itself, where in a few thousand years it would be destroyed or buried. They would place it on the almost unchanging Moon.

And they would have a second reason for doing this. To quote from the original story:

They would be interested in our civilization only if we proved our fitness to survive—by crossing space and so escaping from the Earth, our cradle. That is the challenge that all intelligent races must meet, sooner or later. It is a double challenge, for it depends in turn upon the conquest of atomic energy, and the last choice between life and death.

Once we had passed that crisis, it was only a matter of time before we found the beacon and forced it open. . . . Now we have broken the glass of the fire-alarm, and have nothing to do but to wait. . . .

This, then, was the idea which I suggested in my reply to Stanley Kubrick as the take-off point for a movie. The finding—and triggering—of an intelligence detector, buried on the Moon aeons ago, would give all the excuse we needed for the exploration of the Universe.

By a fortunate coincidence, I was due in New York almost immediately, to complete work on the Time-Life Science Library's *Man and Space,* the main text of which I had written in Colombo. On my way through London I had the first chance of seeing *Dr. Strangelove,* and was happy to find that it lived up to the reviews. Its impressive technical virtuosity certainly augured well for still more ambitious projects.

It was strange, being back in New York after several years of living in the tropical paradise of Ceylon. Commuting—even if only for three stations on the IRT—was an exotic novelty, after my humdrum existence among elephants, coral reefs, monsoons, and sunken treasure ships. The strange cries, cheerful smiling faces, and unfailingly courteous manners of the Manhattanites as they went about their affairs were a continual source of fascination; so were the comfortable trains whispering quietly through the spotless subway stations, the advertisements (often charmingly adorned by amateur artists) for such outlandish products as Levy's bread, the New York *Post,* Piel's beer, and a dozen fiercely competing brands of oral carcinogens. But you can get used to anything in time, and after a while (about fifteen minutes) the glamour faded.

My work in the Time-Life Book Division was not exactly onerous, since the manuscript was in good shape and whenever one of the researchers asked me, "What is your authority for this statement?" I would look at her firmly and reply, "*I* am." So while *Man and Space* progressed fairly smoothly thirty-two floors above the Avenue of the Americas, I had ample energy for moonlighting with Stanley Kubrick.

Our first meeting took place at Trader Vic's, in the Plaza Hotel. The date—April 22, 1964—happened to coincide with the opening of the ill-starred New York World's Fair, which might or might not be regarded as an unfavorable omen. Stanley arrived on time, and turned out to be a rather quiet, average-height New Yorker (to be specific, Bronxian) with none of the idiosyncrasies one associates with major Hollywood movie directors, largely as a result of Hollywood movies. (It must be admitted that he has since grown a full-fledged beard, which is one of his few concessions to modern orthodoxy.) He had a night-person pallor, and one of our minor problems was that he functions best in the small hours of the morning, whereas I believe that no sane person is awake after 10 P.M. and no law-abiding one after midnight. The late Peter George, whose novel *Red Alert* formed the basis of *Dr. Strangelove,* once told me that Stanley used to phone him up for discussions at 4 A.M., desisting only when his bleary-eyed collaborator threatened to retreat to England. I am glad to say that he never tried this on me; in fact I would put consideration for other people as one of his most engaging characteristics—though this does not stop him from being absolutely inflexible once he has decided on some course of action. Tears, hysterics, flattery, sulks, threats of lawsuits will not deflect him one millimeter. I have tried them all: well, most of them. . . .

Another characteristic that struck me at once was that of pure intelligence; Kubrick grasps new ideas, however complex, almost instantly. He also appears to be interested in practically everything; the fact that he never came near entering college, and had a less-than-distinguished high-school career, is a sad comment on the American educational system.

On our first day together, we talked for eight solid hours about science fiction, *Dr. Strangelove,* flying saucers, politics, the space program, Senator Goldwater—and, of course, the projected next movie.

For the next month, we met and talked on an average of five hours a day—at Stanley's apartment, in restaurants and automats, movie houses and art galleries. Besides talking endlessly, we had a look at the competition. In my opinion there have been a number of good—or at least interesting—science-fiction movies in the past. They include, for example, the Pal-Heinlein *Destination Moon, The War of the Worlds, The Day the Earth Stood Still, The Thing,* and *Forbidden Planet.* However, my affection for the genre perhaps caused me to make greater allowances than Stanley, who was highly critical of everything we screened. After I had pressed him to view H. G. Wells's 1936 classic *Things to Come,* he exclaimed in anguish, "What *are* you trying to do to me? I'll never see anything you recommend again!"

Eventually, the shape of the movie began to emerge from the fog of words. It would be based on "The Sentinel" and five of my other short stories of space exploration; our private title for the project was "How the Solar System Was Won." What we had in mind was a kind of semidocumentary about the first pioneering days of the new frontier; though we soon left that concept far behind, it still seems quite a good idea. Later, I had the quaint experience of buying back—at a nominal fee—my unused stories from Stanley.

Stanley calculated that the whole project, from starting the script to the release of the movie, would take about two years, and I reluctantly postponed my return to Ceylon—at least until a treatment had been worked out. We shook hands on the deal during the evening of May 17, 1964, went out onto the penthouse veranda to relax—and at 9 P.M. saw, sailing high above Manhattan, the most spectacular of the dozen UFOs I've observed during the last twenty years.

It was also the only one I was not able to identify fairly quickly, which put me on the spot as I'd tried to convince Stanley that the wretched things had nothing to do with space. *This* one looked exactly like an unusually brilliant satellite; however, the *New York Times*'s regular listing gave no transit at 9 P.M.—and, much more alarming, we felt convinced that this object came to rest at the zenith and remained poised vertically above the city for the best part of a minute before slowly sinking down into the north.

I can still remember, rather sheepishly, my feelings of awe and excitement—and also the thought that flashed through my mind: "This is altogether too much of a coincidence. *They* are out to stop us from making this movie."

What to do? When our nerves had ceased jangling, I argued that there must be a simple explanation, but couldn't think of one. We were reluctant to approach the Air Force, which was still smarting from *Strangelove* and could hardly be blamed if it regarded a report by two such dubious characters as a gag or a publicity stunt. But there was no alternative, so we apologetically contacted the Pentagon and had even gone to the trouble of filling in the standard sighting form—when the whole affair fizzled out.

My friends at the Hayden Planetarium set their computer to work, and discovered that we had indeed observed an Echo 1 transit. Why this spectacular appearance wasn't listed in the *Times*, which gave two later and less impressive ones for the same night, was the only real mystery involved. The illusion that the object had hovered at the zenith almost certainly resulted from the absence of reference points in the brilliantly moonlit sky.

Of course, if it had been a *real* flying saucer, there would have been no movie. Some time later, Stanley tried to insure M-G-M against this eventuality with Lloyd's of London, asking them to draw up a policy which would compensate him if extraterrestrial life was discovered and our plot was demolished. How the underwriters managed to compute the premium I can't imagine, but the figure they quoted was appropriately astronomical and the project was dropped. Stanley decided to take his chances with the Universe.

This was typical of Stanley's ability to worry about possibilities no one else would think of. He always acts on the assumption that if something *can* go wrong, it will; ditto if it can't. There was a time, as the Mariner 4 space probe approached Mars, when he kept worrying about alternative story lines—just in case signs of life were discovered on the red planet. But I refused to cross that bridge until we came to it; whatever happened, I argued, we would be in fine shape. If there *were* Martians, we could work them in somehow—and the publicity for the movie would be simply wonderful.

Once the contracts had been signed, the actual writing took place in a manner which must be unusual, and may be unprecedented. Stanley hates movie scripts; like D. W. Griffith, I think he would prefer to work without one, if it were possible. But he had to have *something* to show M-G-M what they were buying; so he proposed that we sit down and first write the story as a complete novel. Though I had never collaborated with anyone before in this way, the idea suited me fine.

Stanley installed me, with electric typewriter, in his Central Park West office, but after one day I retreated to my natural environment in the Hotel Chelsea, where I could draw inspiration from the company of Arthur Miller, Allen Ginsberg, Andy Warhol, and William Burroughs—not to mention the restless shades of Dylan and Brendan. Every other day Stanley and I would get together and compare notes; during this period we went down endless blind alleys and threw away tens of thousands of words. The scope of the story steadily expanded, both in time and space.

During this period, the project had various changes of title: it was first announced as "Journey Beyond the Stars"—which I always disliked because there have been so many movie Voyages and Journeys that confusion would be inevitable. Indeed, *Fantastic Voyage* was coming up shortly, and Salvador Dali had been disporting himself in a Fifth Avenue window to promote it. When I mentioned this to Stanley, he said, "Don't worry—we've already booked a window for you." Perhaps luckily, I never took him up on this.

The merging of our streams of thought was so effective that, after this lapse of time, I am no longer sure who originated what ideas; we finally agreed that Stanley should have prime billing for the screenplay, while only my name would appear on the novel. Only the germ of the "Sentinel" concept is now left; the story as it exists today is entirely new—in fact, Stanley was still making major changes at a very late stage in the actual shooting.

Our brainstorming sessions usually took place in the Kubrick Eastside penthouse off Lexington, presided over by Stanley's charming artist wife Christiane, whom he met while making *Paths of Glory*. (She appears in its moving final scene—the only woman in the entire film.) Underfoot much of the time were the three— it often seemed more—Kubrick daughters, whom Stanley is in the process of spoiling. Very much of a family man, he has little social life and begrudges all time not devoted to his home or his work.

He is also a gadget lover, being surrounded by tape recorders and cameras—all of which are well used. I doubt if even the most trigger-happy amateur photographer takes as many snapshots of his children as does Stanley—usually with a Pen D half-frame camera, which makes a slight contrast to the Cinerama–Panavision 70-millimeter monster he is maneuvering most of the day. This would seem to suggest that he has no hobbies; it might be more true to say that they are integrated into his work.

He certainly has one absorbing recreation—chess, which he plays brilliantly; for a while he made a modest living at it, challenging the pros in Washington Square. Very fortunately, I long ago decided not even to learn the rules of this seductive game; I was afraid of what might happen if I did. This was very wise of me, for if we had both been chess players I doubt if *2001* would ever have been completed. I am not a good loser.

The first version of the novel was finished on December 24, 1964; I never imagined that two Christmases later we would *still* be polishing the manuscript, amid mounting screams of protest from publishers and agents.

But the first version, incomplete and undeveloped though it was, allowed Stanley to set up the deal. Through 1965, he gathered around him the armies of artists, technicians, actors, accountants, and secretaries without whom no movie can be made; in this case, there were endless additional complications, as we also needed scientific advisers, engineers, genuine space hardware, and whole libraries of reference material. Everything was accumulated during the year at M-G-M's Borehamwood Studios, some fifteen miles north of London; the largest set of all, however, had to be built just six miles south of the city, at Shepperton-on-Thames.

Seventy years earlier, in the twelfth chapter of his brilliant novel *The War of the Worlds,* H. G. Wells's Martians had destroyed Shepperton with their heat ray. *This* year, man had obtained his first close-ups of Mars, via Mariner 4. As I watched our astronauts making their way over the lunar surface toward the ominously looming bulk of the Sentinel, while Stanley directed them through the radios in their space suits, I remembered that within five years, at the most, men would *really* be walking on the Moon.

Fiction and fact were indeed becoming hard to disentangle. I hope that in *2001: A Space Odyssey* Stanley and I have added to the confusion, but in a constructive and responsible fashion. For what we are trying to create is a realistic myth—and we may well have to wait until the year 2001 itself to see how successful we have been.

Postscript

The preceding article was written while *2001* was still in production, when no one—not even Stanley—knew whether we were creating a masterpiece or a disaster, and the release date had been postponed so many times that some feared the title might have to be changed to *2002*.

The article that follows is the only one I have written (or intend to write) *after* the release of the film. It was done at the request of my old friend, and first professional editor, Walter Gillings, for the initial issue of his (alas) short-lived magazine *Cosmos* (April 1969).

The Myth of 2001

23

After five years largely devoted to this single project, I still find myself much too close to it to look at it very objectively. Also, it is now obvious that there is far more in *2001* than I realized when we were making it; perhaps more indeed, than even Stanley Kubrick, its principal creator, had intended.

It is true that we set out with the deliberate intention of creating a myth. (The Odyssean parallel was clear in our minds from the very beginning, long before the title of the film was chosen.) A myth has many elements, including religious ones. Quite early in the game I went around saying, not very loudly, "M-G-M doesn't know this yet, but they're paying for the first $10,000,000 religious movie." Nevertheless, it is still quite a surprise to see how many people realized this, and it has been amusing to see how many faiths have tried to stake claims in the finished work. Several reviewers have seen a cross in some of the astronomical scenes; this is purely a matter of camera composition. I might also mention that we have recently discovered—this was quite a shock—that there is a Buddhist sect which worships a large, black, rectangular slab! The analogy of the Kaaba has also been mentioned; though I certainly never had it in mind at the time, the fact that the Black Stone sacred to the Muslims is reputed to be a meteorite is a more than interesting coincidence.

All the mythical elements in the film—intentional and otherwise—help to explain the extraordinarily powerful responses that it has evoked from audiences and reviewers. In this we have been successful beyond our wildest dreams—certainly beyond mine! I have now read hundreds of reviews from newspapers and magazines all over the world (the most important of these, together with much other material, have appeared in New American Library's *The Making of Kubrick's 2001,* edited by Jerome B. Agel), and a pretty clear pattern of critical reaction is emerging.

A small number of reviewers said, even at first screening, that the movie was a masterpiece and a landmark in the history of the cinema. (Some have remarked flatly that it is "obviously" one of the most important movies ever made.) Another small but significant proportion didn't like it the first time, wrote rather critical reviews, brooded for some days, went to see it again, and then wrote second reviews which were not only recantations but sometimes raves. This is the typical reaction to a new and revolutionary work of art (*vide* the first performance of *The Rite of Spring*), but in the past this process of evaluation took years or decades. I remember saying to Kubrick that he was luckier than Melville, who never lived to see the world appreciate *Moby Dick.*

Moby Dick, of course, has been mentioned many times in connection with *2001;* though it is asking for trouble to make such comparisons, I had this work consciously in mind as a prototype (viz., the use of hard technology to construct a launch pad for metaphysical speculations). It took about half a century before literary criticism caught up with Melville; I wonder how many college theses are now being written on *2001.*

Perhaps the majority of reviews were favorable but somewhat baffled, while another minority group was vociferously hostile. But this very hostility proves the emotional impact of the film; that acute critic Damon Knight (who has written that *2001* is "undoubtedly one of the best films ever made") considers that the extraordinarily obtuse reaction of some science-fiction critics was simply due to embarrassment. They just couldn't face the film's religious implications.

There are others who, quite understandably, expected an updated *Destination Moon* and were baffled by Kubrick's version. But both time and the box office will prove that Kubrick was perfectly correct (indeed, the latter has already done so, for in almost all countries the film has been a fantastic commercial success). To have done a straightforward documentary-type movie—at the very moment when men were preparing to land on the Moon!—would have been to invite disaster, and would have provided no sort of artistic challenge. George Pal's *Destination Moon* was magnificent for 1950; we were interested in starting where that finished.

Soon after the movie was released, and the first cries of bafflement were being heard in the land, I made a remark that horrified the M-G-M top brass. "If you understand *2001* on the first viewing," I stated, "we will have failed." I still stand by this remark, which does not mean that one can't *enjoy* the movie completely the first time around. What I meant was, of course, that because we were dealing with the mystery of the Universe, and with powers and forces greater than man's comprehension, then by definition they could not be totally understandable. Yet there is at least one logical structure—and sometimes more than one—behind everything that happens on the screen in *2001,* and the ending does not consist of random enigmas, some simple-minded critics to the contrary. (You will find *my* interpretation in the novel; it is not necessarily Kubrick's. Nor is his necessarily the "right" one—whatever that means.)

2001 has already become part of film history; it is the first science-fiction movie to do so, and its success has been so overwhelming that it poses the embarrassing problem "Where do we go from here?" in a particularly acute form. Yet in a very few years it will probably seem old-fashioned, and people will wonder what all the fuss was about.

As for the dwindling minority who *still* don't like it, that's their problem, not ours. Stanley and I are laughing all the way to the bank.

Postscript

A much fuller account of the making of the film, together with material which was never used in the final version, will be found in *The Lost Worlds of 2001* (New York: New American Library, 1972).

About the Author

Arthur C. Clarke was born in Somerset, England. He served in the Royal Air Force in World War II, becoming technical officer in charge of the first experimental Ground-Controlled Approach unit. Working for a time with the American scientists who had developed this equipment, he later assumed command for the RAF. He is a fellow of the Royal Astronomical Society and served as chairman of the British Interplanetary Society for some years.

Mr. Clarke has written many books in the field of space. His *Interplanetary Flight* was a pioneer in the literature of space flight. Other popular titles about space and the modern age are *The Promise of Space, Profiles of the Future,* and *Voices from the Sky.* He has also written *The Coast of Coral,* about skin diving at the Great Barrier Reef, many books for young people, and science fiction. He spends much of his time in Ceylon.

In September, 1962, Clarke was awarded the Kalinga Prize in New Delhi and, in October, 1963, the Stuart Ballantine Medal of the Franklin Institute "for his soundly based and prophetic early concept of the application of satellites in the primary human endeavor of communication."

He has written with Stanley Kubrick the M-G-M film *2001: A Space Odyssey.*